Stalking the Riemann Hypothesis

Stalking the Riemann Hypothesis

The Quest to Find the Hidden Law of Prime Numbers

Dan Rockmore

Pantheon Books  *New York*

 Published in the United States by Pantheon Books, a division of Random House, Inc., New York, and simultaneously in Canada by Random House of Canada Limited, Toronto.

Pantheon Books and colophon are registered trademarks of Random House, Inc.

Library of Congress Cataloging-in-Publication Data
Rockmore, Daniel N. (Daniel Nahum)
Stalking the Riemann hypothesis : the quest to find the hidden law of prime numbers/Dan Rockmore.
p. cm.
Includes index.
ISBN 0-375-42136-X
1. Numbers, Prime. 2. Number theory. 3. Riemann, Bernhard, 1826–1866. I. Title.
QA246.R63 2005
512.7′23—dc22 2004053525

www.pantheonbooks.com

Printed in the United States of America

First Edition

1 2 3 4 5 6 7 8 9

For Loren Butler Feffer (1962–2003),
scholar, athlete, and dear friend

Contents

Acknowledgments

The science writer faces an interesting intellectual challenge. To distill years, even centuries, of scientific investigation for a broad and curious audience, while not raising the hackles of the experts in the field, is something of an intellectual tightrope walk. It is my hope that in this book I haven't fallen off the wire too many times, and I'm grateful to the family, friends, and colleagues who have either steadied me during this performance, or at least allowed me a soft landing from time to time. I'd like to take a moment to thank some of them.

I have been lucky to have had comfortable circumstances in which to write, and I appreciate the hospitality provided to me by New York University's Courant Institute, the Institute for Advanced Study, the Santa Fe Institute, and of course my home turf: Dartmouth College's departments of mathematics and computer science. I would like to give special thanks to the Institute for Advanced Study's librarian Momota Ganguli for her help in tracking down a few hard-to-find references, as well as Dartmouth College's librarian Joy Weale for the quick attention she gave to my many requests for materials.

Katinka Matson helped me find a home for this book with Random House, where Marty Asher, Edward Kastenmeier, Will Sulkin, and Eric Martinez have worked hard to steer it through the shoals of the editorial and production process. Special thanks to Susan Gamer for a terrific job of copyediting.

I am grateful to the many people who read portions of the book in various stages of completion. In particular, I'd like to thank Percy

Deift, Freeman Dyson, Cormac McCarthy, Peter Sarnak, Craig Tracy, Harold Widom, and Peter Woit for their helpful comments and corrections, Eric Heller and Jos Leys for furnishing some key figures, and Andrew Odlyzko for providing easy access to the zeta zero data. Sincere thanks to Peter Doyle for translating the relevant pieces of the Stieltjes-Hermite correspondence, and for keeping local interest alive through his leadership of the always entertaining "Riemann Seminar" at Dartmouth. Peter Kostelec, a terrific scientist and friend, made many of the figures in this book, as well as many more that weren't used. Bob Drake, resident "folk mathematician," gave me many good comments on early versions of the manuscript and many more good squash lessons. Thanks to Laurie Snell for his close reading of an early manuscript and for his constant encouragement. Greg Leibon's careful attention helped me avoid a few bloopers at the end. Of course, I take full responsibility for any errors that made it past the eagle eyes of my vigilant friends and colleagues.

My parents, Ron and Miriam; my brother Adam; my sister-in-law Alicia; and my delightful niece Lucy always have given me the luxury and benefit of a loving and supportive family. If it is true that the apple doesn't fall too far from the tree, I am lucky to have come from sturdy stock giving generous shade.

Ellen—my wife, love, confidante, and super-editor—and our son, Alex, have brought to my life a light and joy that I had not thought possible. I'm grateful to them every day. Through every stage of this project, my dear and now departed dog Digger was a source of constant comfort and calm. I always appreciated and now miss his nuzzling reminders of the necessity of regular breaks—in good weather and bad. If there is a sonorous sentence or two in this book, it was almost surely rocked loose by the steady footfall of one of our countless but countable walks. I like to think that his playful and gentle spirit lives on in these pages.

Finally, this book is dedicated to the memory of my dear friend Loren Butler Feffer, whose keen mind, boisterous good humor, and fierce loyalty helped me through many a difficult time, both personal and professional. She is ever missed. Loren—I wish you could have been there when we popped open the champagne. This one's for you.

Stalking the Riemann Hypothesis

1 *Prologue—It All Begins with Zero*

It's one of those slate-gray summer days that more properly belong to mid-August than late May, one of those days in New York City when it is barely clear where the city ends and the sky begins. The hard-edged lines and Euclidean-inspired shapes that are building, sidewalk, and pavement all seem to fuse into one huge melted mass that slowly dissolves into the humid, breezeless, torpid air. On mornings like this, even this irrepressible metropolis seems to have slowed a notch, a muffled cacophony more bass than treble, as the city that never sleeps stumbles and shuffles to work.

But here in Greenwich Village, at the corner of Mercer and West Fourth streets, where we find New York University's Warren Weaver Hall, the hazy torpor is interrupted by a localized high-energy eddy. Here, deep in the heart of the artistic rain forest that is "the Village," just across the street from the rock 'n' rolling nightclub the Bottom Line, a stone's throw from the lofts and galleries that gave birth to Jackson Pollock, Andy Warhol, and the Velvet Underground, is the home of the Courant Institute of Mathematical Sciences, where at this moment there is an excitement worthy of any gallery opening in SoHo, or any new wave, next wave, or crest-of-the-wave musical performance.

The lobby and adjacent plaza are teeming with mathematicians, a polyglot and international group, abuzz with excitement. Listen closely, and amid the multilingual, every-accent mathematical jibber-jabber you'll hear a lot of talk about nothing, or more properly a lot of talk about zero.

Zero is not an uncommon topic of conversation in New York, but more often than not it's the "placeholder zeros" that are on the tip of the New Yorker's tongue. These are the zeros that stand in for the orders of magnitude by which we measure the intellectual, cultural, and financial abundance that is New York: one zero to mark the tens of ethnic neighborhoods, two for the hundreds of entertainment options, three for the thousands of restaurants, six for the millions of people, and, of course, the zeros upon zeros that mark the billions or even trillions of dollars that churn through the city every day. These are not the zeros of void, but the zeros of plenty.

But, today, just one week past Memorial Day 2002, it's a zero of a different flavor which has attracted this eclectic group to downtown New York City. Here some of the world's greatest mathematicians are meeting to discuss and possibly, just possibly, witness the resolution of the most important unsolved problem in mathematics, a problem that holds the key to understanding the basic mathematical elements that are the prime numbers. The zeros that tip the tongues of these mathematical adventurers are ***zeta zeros,**** and the air is electric with the feeling that perhaps this will be the day when we lay to rest the mystery of these zeros, which constitutes the ***Riemann hypothesis.***

For over a century mathematicians have been trying to prove the Riemann hypothesis: that is, to settle once and for all a gently asserted conjecture of Bernhard Riemann (1826–1866), who was a professor of mathematics at the University of Göttingen in Germany. Riemann is perhaps best known as the mathematician responsible for inventing the geometrical ideas upon which Einstein built his theory of general relativity. But in 1859, for one brief moment, Riemann turned his attention to a study of the long-familiar prime numbers. These are numbers like two, three, five, and seven, each divisible only by one and itself, fundamental numerical elements characterized by their irreducibility. Riemann took up the age-old problem of trying to find a rule which would explain the way in which prime numbers are distributed among the whole numbers, indivisible stars scattered without end throughout a boundless numerical universe.

In a terse eight-page "memoir" delivered upon the occasion of his

*Terms in ***boldface italic*** can be found in the glossary.

induction into the prestigious Berlin Academy, Riemann would revolutionize the way in which future mathematicians would henceforth study the primes. He did this by connecting a law of the primes to the understanding of a seemingly completely unrelated complex collection of numbers—numbers characterized by their common behavior under a sequence of mathematical transformations that add up to the *Riemann zeta function.* Like a Rube Goldbergesque piece of mathematical machinery, Riemann's zeta function takes in a number as raw material and subjects it to a complicated sequence of mathematical operations that results in the production of a new number. The relation of input to output for Riemann's zeta function is one of the most studied processes in all of mathematics. This attention is largely due to Riemann's surprising and mysterious discovery that the numbers which seem to hold the key to understanding the primes are precisely the somethings which Riemann's zeta function turns into nothing, those inputs into Riemann's number cruncher that cause the production of the number zero. These are the zeta zeros, or more precisely the zeros of Riemann's zeta function, and they are the zeros that have attracted a stellar cast of mathematicians to New York.

In his memoir, Riemann had included, almost as an aside, that it seemed "highly likely" that the zeta zeros have a particularly beautiful and simple geometric description. This offhand remark, born of genius and supported by experiment, is the Riemann hypothesis. It exchanges the confused jumble of the primes for the clarity of geometry, by proposing that a graphical description of the accumulation of the primes has a beautiful and surprisingly simple and precise shape. The resolution of Riemann's hypothesis holds a final key to our understanding of the primes.

We'll never know if Riemann had in mind a proof for this assertion. Soon after his brief moment of public glory, the ravages of tuberculosis began to take their toll on his health, leaving him too weak to work with the intensity necessary to tie up the loose ends of his Berlin memoir. Just eight years later, at the all too young age of thirty-nine, Riemann was dead, cheated of the opportunity to settle his conjecture.

Since then, this puzzling piece of Riemann's legacy has stumped the greatest mathematical minds, but in recent years frustration has begun to give way to excitement, for the pursuit of the Riemann hypothesis has begun to reveal astounding connections among nuclear physics, chaos, and number theory. This unforeseen confluence of mathematics and physics, as well as certainty and uncertainty, is creating a frenzy of activity that suggests that after almost 150 years, the hunt might be over.

This is the source of the buzz filling the Courant Institute's entryway. It is a buzz amplified by the fact that whoever settles the question of the zeta zeros can expect to acquire several new zeros of his or her own, in the form of a reward offered by the Clay Institute of Mathematics, which has included the Riemann hypothesis as one of seven "Millennium Prize Problems," each worth $1 million. So the jungle of abstractions that is mathematics is now full of hungry hunters. They are out stalking big game—the resolution of the Riemann hypothesis—and it seems to be in their sights.

The Riemann hypothesis stands in relation to modern mathematics as New York City stands to the modern world, a crossroads and nexus for many leading figures and concepts, rich in unexpected and serendipitous conjunctions. The story of the quest to settle the Riemann hypothesis is one of scientific exploration and discovery. It is peopled with starry-eyed dreamers and moody aesthetes, gregarious cheerleaders and solitary hermits, cool calculators and wild-eyed visionaries. It crisscrosses the Western world and includes Nobel laureates and Fields medalists. It has similarities with other great scientific journeys but also has its own singular hallmarks, peculiar to the fascinating world of mathematics, a subject that has intrigued humankind since the beginning of thought.

2 *The God-Given Natural Numbers*

THE GREAT GERMAN mathematician Leopold Kronecker (1823–1891) said that "God created the natural numbers." And it is true that the ***natural numbers***—one, two, three, four; on and on they go—appear to have been present from the beginning, coming into existence with the birth of the universe, part and parcel of the original material from which was knit the ever-expanding continuum of space-time.

The natural numbers are implicit in the journey of life, which is a nesting of cycles imposed upon cycles, wheels within wheels. One is the instant. Two is the breathing in and out of our lungs, or the beat of our hearts. The moon waxes and wanes; the tides ebb and flow. Day follows night, which in turn is followed once again by day. The cycle of sunrise, noon, and sunset give us three. Four describes the circle of seasons.

These natural numbers help us to make sense of the world by finding order, in this case an order of temporal patterns, that lets us know what to expect and when. We notice the rising and setting of the sun, and that cycle of two is given a more detailed structure as we follow the sun through the sky over the course of a day. We turn the temporal telescope around and also see day as part of the larger cycle of the phases of the moon, whose steady progress is situated within the cycle of the seasons that makes up the year. Patterns within patterns within patterns; numbers within numbers within numbers—all working together to create a celestial symphony of time.

Armed with this new understanding we make tentative, tiny forays out into the Jamesian "booming, buzzing world" and shape a life within and around it. Embedded in the recognition of the cycle is the ability to predict, and thus to prepare, and then to direct the world to our advantage. We coax and bend an unflinching, steady march of time; and in a subtle jujitsu of nature, technologies are born. We learn when to sow and when to reap, when to hunt and when to huddle. We exploit that which we cannot change. We discover the cycle and ride it as an eagle rides an updraft.

In the absence of a natural cycle we may impose one, for in routine we find a sense of control over the unwieldy mess that is life. We relish the comfort of being a regular at a local diner or a familiar face at the coffee cart on the street, and the rhythm of the daily morning dog walk. We dream of options, if only to choose our own routines, our own patterns, our own numbers.

But as befits that which is part and parcel of space-time, number is not only a synecdoche of temporal organization but also the most basic and elementary means of quantifying a spatial organization of the world. Nature gives us few, if any, truly straight lines or perfect circles. But there is one moon; there is one sun; the animals go two by two. We organize, we count, and therefore we are.

In this way, number is presented to us in the world in both time and space, instances upon instances, but this is only the beginning. Kronecker said not only that the natural numbers are God-given, but also that "all else is the work of man." What first appear as singular phenomena are eventually unified, gathered into a collective that is then recognized as a pattern. Soon, the pattern is itself familiar, and so it becomes less a pattern and more a particular. The game is then repeated, and we find a new superpattern to explain what had once seemed disparate patterns. So on and so on we go, building the discipline that will come to be known as mathematics.

Beginning "the work"

Suppose that I walk past a restaurant and catch a glimpse of a perfectly set square table, place settings at each edge, each side of the table providing a resting place for a full complement of plates, glasses, and silverware. As I approach the entrance to the restaurant, a group of women arrive and I imagine them seated at that beautiful table, one at each side, continuing their animated conversation. As I pass by again some time later, I see the women leave the restaurant. They stand outside, say their good-byes, and one by one are whisked away by taxicabs.

What is it that the group of women has in common with the collection of place settings, the chairs at the table, the very sides of the table, and the taxicabs that finally take them away? It is the correspondence that they engender. It is a correspondence that I make mentally and visually as I watch the women, one that you make as you read this story, seeing each woman paired with a chair, a plate, or a taxi. Any other grouping of objects that could be paired with them in this way has this same property, this same basic pattern. This pattern is one of "numerosity." These groups all share the property of "fourness."

Each collection, whether it be the chairs, the place settings, or the taxis, is such that its component objects can be put into a one-to-one correspondence with the group of women. We say, as an abbreviation for this property, that the group of women has a size of four, and this is a property shared among each of the sets of objects that may be put into a correspondence with the women. If you had in your possession a collection of hats and I inquired if you had one for each of the women, you might have me list the women, or show you a picture of the group, but even better, you could ask me "how many" women need hats. My answer, "four," would be enough for you to check to see if you had one hat for each.

The self-contained nature of the correspondence—there is no object left unpaired—is perhaps what underlies the other classification of the number four, or for that matter any natural number, as an ***integer,*** and in particular a positive integer. The totality of the integers consists of the natural numbers, their negatives, and zero.

Thus four becomes an agreed-upon name for a pattern that we recognize in the world. At Christmas, four are the calling birds; at Passover, four are the matriarchs, symbols that are simultaneously iconic and generic. We wind our way back through numerical history. Four are the fingers proudly displayed by a protonumerate toddler, a set of scratchings on a Sumerian cuneiform, or the bunch of beads or pebbles lying at the feet of a Greek philosopher. The last of these universal physical numerical proxies, which the Greeks called *calculi,* gave birth to our words ***calculus*** and *calculate,* and mark the mathematician as both the forefather and the child of the first "bean counters": the Pythagoreans.

The Pythagoreans, followers of the great Greek mathematician Pythagoras (580–500 B.C.E.), are said to have first abstracted the particular instances of a number to a universal and generic pile of pebbles or arrangements of dots in the sand. For the Pythagoreans, mathematics and the mathematical were ubiquitous. They lived by the motto "Number is all," and as a consequence numerology was integral to their daily lives. As for Pythagoras, very little is known of his life. He was a mystic and scholar who led a cult whose basic tenet was devotion to the study of knowledge for its own sake. To describe this pursuit, Pythagoras coined the word *philosophy.* As applied to the specific study of number and geometry, this sort of investigative approach led him to create another discipline, which he named *mathematics,* meaning "that which is learned."

Before the Pythagoreans, mathematical studies were by and large contextualized through their particular applications. Records are sketchy, but certainly the Egyptians and Babylonians developed a great deal of basic mathematical (or, perhaps more accurately, "calculational") procedures for dealing with the everyday problems of their agricultural, architectural, or commercial activities. The mathematical investigations of the Pythagoreans appear to mark the origins of a pure study of number, as opposed to a study of number in the context of its potential or actual applications. So Pythagoras, the "father of mathematics," gave birth to a poetry or art of pattern, an intellectual discipline in which numbers themselves were studied as arrangements of beautiful, regular, rhythmic shapes. Herein lies the origin of a study of the laws that govern patterns within the patterns or the behavior of numbers. This is ***number theory,*** and the Pythagoreans are history's first ***number theorists.***

The simplest patterns to be found were those of the geometric shapes in which only certain numbers of pebbles could be arranged. There were *square* numbers, such as one, four, nine, and sixteen, and *triangular* numbers such as one, three, and six,

and even *hexagonal* numbers like one and six,

each providing an infinite family of dimpled patterns ready for categorization and study. These few examples represent but a shallow draw from a potentially infinite well of possibilities. Any of your favorite two-dimensional shapes could yield other classes of *figurate* numbers, and many others were studied.

Primes: the fundamental pattern within the pattern

The Pythagoreans plumbed the piles of pebbles in search of structure and regularity, and so might I also look for patterns as I pass by my little imagined restaurant. One evening I peer though the window and notice that the room is full. How many diners are there on this busy evening? I could count them all one by one if I stood there long enough, but I find that I can't help trying to take a shortcut. The tables organize the diners into groups of four, and so I know that the number of diners will simply be four times the number of tables. The tables are arranged tastefully and delicately about the softly lit room, and again my eye looks for patterns. I first cluster the tables in pairs and notice that this leaves one table by itself. Groupings of three leave two tables adrift. Groupings of four again leave a lone solitary dinner party, and so on. My search for an even regularity in the arrangement of tables proves fruitless, and finally I surrender, forced to enumerate them. Seventeen tables, making seventeen parties of four, and thus, sixty-eight diners in all.

Seventeen. A curious number. No matter how I try to organize the tables into a pattern of similar patterns, into clusters of equinumerous clusters, I am foiled. One more or one less, and the reduction is possible. Sixteen tables could yield four groupings of four tables, or two groupings of eight tables. Eighteen tables would give three groupings of six, or two groupings of nine. Thus sixteen and eighteen are a composition of other, smaller numbers: ***composite,*** we might say. Seventeen, however, is not a number of numbers, it is not a pattern of patterns. It is a first-order pattern, a primal pattern, a ***prime*** pattern.

Primes and composites. It appears that the Greeks were the first to be aware of this distinction of the natural numbers. Perhaps it was a thoughtful shepherd clustering his sheep, or an acquisitive merchant counting out bolts of cloth, sheaves of wheat, or jugs of wine, who first discovered this distinction. The mystical mathematical musings of the Pythagoreans included a taxonomy of number dependent upon the concept of ***divisibility,*** whereby a given number can be obtained as some multiple of a specific number. Twelve is divisible by three, but not by five. The Pythagoreans called a number *perfect, abundant* (in the sense of having many divisors), or *deficient* according to its property of being (respectively) equal to, less than, or greater than the sum of its divisors. For example, twenty-eight is perfect because it is the sum of its divisors: one, two, four, seven, and fourteen. The number twelve falls short of the sum of one, two, three, four, and six, and is thus said to be abundant, and eight, which exceeds the sum of one, two, and four, is deficient. *Amicable* numbers are two numbers defined by the property that the sum of divisors of the one is equal to the other. An example of such a pair is 220 and 284. Although the Pythagoreans were intrigued by the concept of divisibility, the first written record of classification into primes and composites dates from about 200 years after the death of Pythagoras and is found in the mother of all mathematical textbooks, Euclid's *Elements.*

As with Pythagoras, the exact origins of Euclid (c. 330–275 B.C.E.) remain a mystery, but it is presumed that he was a student of Plato, or at least attended Plato's Academy, a scholarly retreat whose gateway bore the famous inscription "Let no man ignorant of geometry enter here." He then seems to have been called to Alexandria to teach mathematics at "the Museum," an institution of higher learning set up by Ptolemy I. Euclid was the author of several texts, but he is surely most famous for the *Elements,* which comprises thirteen books and is today thought of as primarily a textbook in geometry. It follows a largely axiomatic, or *synthetic,* approach to geometry in which "self-evident" geometric truths are laid down as axioms from which all of Euclidean geometry is derived through the application of Aristotlean logic. It is striking how much of *Elements* could today still be read and understood by any attentive schoolchild, and indeed, this axiomatic approach continues to be taught in many schools.

Elements takes a geometric approach to number, whose study is mainly concentrated in Books 7 through 9. The Pythagoreans believed that number is all, but for Euclid and his followers, geometry ruled the universe and only those mathematical ideas that had geometric realizations in terms of the perfection of ideal straight lines and circles (constructions capable of being realized with only the most basic tools of the working geometer: pencil, straightedge, and compass) were of interest or even acknowledged.

In such a mathematics, all investigations begin with a unit length that serves as the basic measuring stick against which all other lengths are compared or measured. These lengths are the geometric realization of numbers. Addition of whole numbers corresponds to joining known lengths end to end. Multiplication of whole numbers is then achieved by repeated addition. It soon becomes evident that not all lengths (whole numbers) can be obtained by multiplying together two smaller numbers: that is, not all numbers are divisible. In this geometric setting, divisibility is expressed as the possibility that one number might be "measured" in terms of another (i.e., that a given length could be "measured" as some number of copies of a given length). Thus, one number divides another if the corresponding length of the one "measures" the other.

Among all lengths (numbers) those that are measured by no others are distinguished as the primes.

PRIMES: THE INTEGRAL ATOMS

So here we have the fundamental dichotomy within the natural numbers: either a natural number is measured by another or it is not; either it is composite or it is prime. Herein lies the atomic nature of the primes, for if a natural number is composite, then we split it, recognizing it as the product of two necessarily smaller natural numbers. We continue the splitting, ultimately arriving at those prime (indeed, primal) pieces incapable of any further division.

To give one simple example, the number 2,695 is split into the product 49 × 55, the former of which can in turn be split into the product 7 × 7, while the latter is the product 5 × 11. The numbers 5, 7, and 11 cannot be split any further.

Thus, given any natural number, by an analogous successive iteration of splittings, continued so long as is possible, a collection of prime numbers is revealed, which if recombined by multiplication will yield the original number. In our example, 2,695 is the product of the primes 5 and 11, and two copies of the prime 7. This is called its ***prime factorization.***

Just as any chemical compound is composed of integral amounts of its constituent elements (exactly two atoms of hydrogen and one of oxygen make water; one sodium atom and one chlorine atom combine to give us a molecule of table salt), so any number is produced as a product of particular primes, thereby yielding an arithmetic molecule. However, whereas it is possible for two chemically distinct molecules to have the same molecular formula (such substances are called molecular isomers), a number's prime factorization is unique. This property of unique factorization is so basic to numbers that it is often called the ***fundamental theorem of arithmetic.***

To uncover a fundamental element, the chemist takes a sample and subjects it to all sorts of tests: dabbing it with acid, holding it to a flame, boiling it in water. Physicists go a step further, looking for constituents that are even more fundamental, by bombarding materials with high-energy particles. What then of mathematics and the search for the primes? How to find these numbers that are measured by no others? How to mine the mountains of number for the irreducible, prime nuggets? It turns out that we need nothing as dramatic as a supercollider. All we need is a simple sieve, the sieve of the Greek mathematician and philosopher Eratosthenes.

In search of primes

Eratosthenes (c. 276–194 B.C.E.) was a native of Cyrene. Like Euclid, he was called to Alexandria, where he ran the library at the university and also served as tutor of the son of Ptolemy III. Eratosthenes is well-known for his early estimate of the circumference of the earth, assuming it to be a perfect sphere. Using only basic geometry and observations of the position of the sun, Eratosthenes arrived at a measurement of about 24,660 miles, a pretty close approximation to the true value (a bit more than 24,901 miles).

The "sieve of Eratosthenes" provides an orderly procedure for culling the primes from the natural numbers. The indivisible nature of the primes implies that once a prime is identified, none of its multiples can be a prime. Since two is a prime, we can remove from consideration all of its multiples—that is, all the even numbers—effectively "sieving" them from the body of natural numbers subject to our search for primes. We then look for our next prime among the first number not removed as a multiple of two, and it is three. Once again, we sift the remaining numbers for multiples of three, none of which are possible primes. Excluding the primes two and three, we now are left with only numbers that are multiples of neither two nor three, the smallest of which must again be prime—and lo and behold, it is: the number five.

We continue in this fashion, repeatedly sifting the integers with sieves of a size that grows as the primes grow. Implicit in Eratosthenes' sieve is one of the first ***primality tests:*** in order to determine whether a given number is prime, see if at any point in the sifting process it is removed from consideration. For the smaller natural numbers this test is useful; but for larger ones, those on the order of hundreds of digits, such a simple-minded test is impossible to conduct. Potentially, even the fastest computers of our day would require more time than the age of the universe to work on such a problem.

Eratosthenes was simply interested in beginning to map the primes, and took a first stab at identifying the primal archipelago dotting the integer sea. Today, our interest in the primes is at least as intense, although perhaps more prosaic, for the primes provide the alphabet as well as the language that both enables and secures digital communication.

SPEAKING IN PRIMES

In any single computer—or over the vast, highly distributed, interconnected network of computers that forms the Internet—resides a self-contained, logical world, whose language and phenomena find a common expression in number. This is the *digital* world, a silent world of a forever flowing stream of zeros and ones, strings of electronic fingers and toes, either bent at the knuckle or extended. These ***binary*** sequences are both its form and its function, providing instruction and carrying information. Zeros and ones make up the alphabet for the language used to describe, as well as to direct, events, providing a means of expression for the multiline program poems of executive orders. They are responsible for communicating our ideas, thoughts, conversations, and business transactions.

The rules for manipulating the ***bits*** (binary digits) which are these zeros and ones are based on the smallest of prime cycles, the cycle of length two. Reasoning is distilled to a logic which becomes arithmetic, capable of translation into complicated circuitry etched onto silicate wafers which serve to mediate the communion of thought and machine. These rules embody the distillation of debate and dialectic. In this stripped-down logic by which our machines live, we see mirrored the basic rhythms of life: breathe in, breathe out; yes, no; true, false; on, off. They are the distillation of us, simplified and pristine.

Although the simplest of primes gives voice to the Internet revolution, it is the larger primes that enable an intelligible and credible discussion. The streams of information that course through the wires and airwaves are processed in huge chunks, whose sizes are determined by their prime factors. In this way the prime decomposition of a number is crucial to the efficient communication and analysis of the images, music, financial information, etc., which reside in our home computers and are sent across the Internet.

Prime numbers are also fundamental to many of the processes that enable the reliable (i.e., error-free) transmission of digitized information. These are primes used in the service of *error correction.* The simplest form of error correction appends a single bit to a binary message, setting it equal to zero if the message contains an even number of ones, and equal to one if the message contains an odd number of ones. That is, a single bit records the remainder obtained upon dividing the total number of ones in the binary message by the prime two. That remainder is called the *parity* of the number of ones. The message is now sent to the receiver, along with the parity bit. If during transmission a single bit of the message is changed, then the parity of the received message will differ from the received parity bit, indicating the message has been corrupted. More elaborate error-correction schemes use arithmetic properties of other primes.

The biggest use of primes is found in the technology of *digital cryptography.* This is the body of techniques used to ensure that information is sent securely over the Internet. In this application, primes of tremendous magnitudes, on the order of hundreds of digits long, are used to *encrypt* messages in such a way that no present (or even foreseeable) technology would have a chance of deciphering them should they be intercepted.

These cryptographic schemes start by interpreting the digital messages as numbers. One way in which this is possible is through the interpretation admitted by any finite list (string) of zeros and ones as a number's ***binary expansion.*** Reading right to left, any binary string represents an accumulation of successive powers of two in which a power is included only if the symbol in the corresponding position is a one. For example, the string 10011 would be interpreted as the sum of one zeroth power of two (which is one), one first power of two, no second or third powers of two, and one fourth power of two. In short:

$$(1 \times 2^0) + (1 \times 2^1) + (0 \times 2^2) + (0 \times 2^3) + (1 \times 2^4)$$

which is the sum of sixteen, two, and one, and gives nineteen. This is akin to our familiar decimal expansion, in which numbers read right to left are shorthand for an accumulation of powers of ten. For example, 219 is shorthand for nine zeroth powers of ten added to one first power of ten and two second powers of ten.

Once a message is interpreted as a number, it can be scrambled by applying any of a variety of mathematical operations. The procedures that ensure the security of most Internet credit card transactions first repeatedly square a number (i.e., computing some very high power of the message-bearing number) and then divide this very high power by a previously specified number and record the remainder. This is done for various powers, and the final encryption of the original message is the number obtained by then multiplying together these numbers and taking one last remainder. The number of times this basic step occurs, as well as the number by which we divide, depends on some previously specified pair of huge prime numbers.

In this way, the primes underlie all of modern digital technology, as the means by which information is made palatable to the computer. They are like the air we breathe—necessary, but superficially invisible, the ghost in the digital machine.

Central to the cryptographic uses of primes is the underlying assumption that they are sufficiently plentiful. Quantification of their plenitude is precisely what the Riemann hypothesis is all about. So, as a first step toward its statement, let's start by trying to count the primes.

EUCLID'S PROOF OF THE INFINITUDE OF PRIMES

We know that there is no largest natural number. Posit such a quantity, and this conjectured *supremum* is defeated by the number obtained by adding one. That's what we mean by infinity. Like two children shouting in turn,

"I dare you!"

"I double-dare you!"

"I triple-dare you!"

the count could go on and on forever. But what of the numerosity of the primes? The totality of natural numbers is infinite, but do the primes need go on and on as well? After all, four simple molecules suffice to create the seemingly infinite diversity of life cast within DNA.

Astrophysicists believe that a finite number of basic particles are responsible for our presumably unbounded universe. Could it be that the primes too are finite? Euclid was the first person to demonstrate that the answer to this question is no.

Euclid proved to the world that the primes are infinite by a clever argument which we might recast as a Socratic dialogue:

Euclid: Dan, I would like to impress upon you that, should we travel down the road of the number line, we might always find a prime number ahead.

Dan: Indeed, wise Euclid, could this be so? How to be sure that no matter how far we travel there is always a prime number ahead?

Euclid: My dear Dan, as I travel along the number line I acquire prime numbers. At any point, I might stop to rest and wonder if indeed I now have them all. I might care to create a new number from the finite collection in hand by multiplying together all of the primes in my possession, and then increasing this already large number by one.

Dan: Yes, Euclid, you certainly could create that number if you like.

Euclid: Now Dan, this large number is, no matter how large, still a number, and as such it must have an elemental factorization in terms of prime numbers. Isn't that so?

Dan: Yes, that is most certainly so.

Euclid: Do we agree that any prime number which is part of the factorization of my large number must divide that number exactly, so that there is no remainder?

Dan: Absolutely, for that is what it means to be a factor.

Euclid: Aha! Here lies the problem. Should I divide my newly constructed number by any of the primes in my possession, we see that we get a remainder not of zero, but of one! So, it cannot be that any of the primes currently in my possession divides our new number.

Dan: Yes, that must be true.

Euclid: Thus, there must be a new prime number out there on the number line, either that number out in the distance that I have just constructed, or another that lies somewhere in between here and there.

Dan: Yes, that must be true.

Let's illustrate Euclid's argument with an example. Suppose we stop our little stroll along the number line at the number thirteen, having identified the primes up to that point: 2, 3, 5, 7, 11, and 13. Euclid suggests that we take a look at the number $(2 \times 3 \times 5 \times 7 \times 11 \times 13) + 1$. A quick calculation reveals that the result is the number 30,031.

The number 30,031 cannot be divisible by any of the primes 2, 3, 5, 7, 11, or 13, because any one of these gives a remainder of one when we attempt to perform the division. For example if we try to divide 7 into $(2 \times 3 \times 5 \times 7 \times 11 \times 13) + 1$ we see that it goes into this number $(2 \times 3 \times 5 \times 11 \times 13)$ times, with a remainder of 1. Nevertheless, as any natural number is either itself prime, or composed of a product of smaller prime factors, it must then be that the number 30,031, while not necessarily prime, is at least made up of primes other than 2, 3, 5, 7, 11, and 13.

Euclid has discovered a procedure that takes in a collection of prime numbers and outputs a new number which is composed of prime numbers that are not among our original collection. Through a sort of alchemy of arithmetic, the prime numbers 2, 3, 5, 7, 11, and 13 are mixed together by multiplication, and the result is incremented by one in order to produce the number 30,031. When this number is subjected to the fires of factorization, two new primes are revealed: we write $30{,}031 = 59 \times 509$, and note that each prime factor is greater than any of the original prime ingredients. In this way, old primes are ground by arithmetic mortar and pestle to create new primes. It is a form of mathematical magic, perpetually producing primes.

3 *The Shape of the Primes*

EUCLID'S INVESTIGATIONS uncovered the fact that there are an infinity of primes. But this simple description sweeps much under the rug. Legend has it that the Inuit have a virtual avalanche of words to describe snow; similarly, mathematicians' familiarity and fascination with infinity have led to an appreciation of infinity as a highly textured and nuanced notion.

To say simply that the primes are infinite is, to a mathematician, not the end of a discussion but a beginning. "How infinite is it?" is a sensible mathematical question, and the way in which mathematics tackles the infinite is among the most distinguishing features of this science.

Once again, we can use the analogy of the study of the universe. We assume it is infinite, but to what extent? Some say the universe is ever-expanding; others say that expansion will eventually give way to a great compression, the Big Bang folding back into the Big Crunch. These and other astrophysical theories depend upon the way in which matter and energy are strewn about the cosmos—the light and the dark, that which we can see and that which we can't. Are matter and energy spread about in such a way that eventually the expansion will halt? Or is there so little of each, so widely scattered, that the universe will grow forever? Where do matter and energy lurk and from where do they arise? Questions of form and questions of extent confront us.

Similarly, we ask questions about the way in which the primes are distributed among the natural numbers. Are they speeding away, ever farther apart, or are they clustered together in recognizable constellations? In what sense are the primes infinite? Implicitly, these are questions of both quantity and quality, or more precisely, of ***cardinality*** and ***asymptotics.*** We'll look at these ideas now.

CARDINALITY: HOW MANY PRIMES ARE THERE?

On the surface, the declaration that the primes are infinite simply means that a list of the primes would go on forever. A first prime of two, a second prime of three, a third prime of five, a fourth prime that is seven, and so on and so on. Both explicit and implicit is the correspondence that we create in our listing of the primes—a correspondence attaching to any prime number that natural number which indicates its position as we list the primes in increasing order.

Any collection of objects (or numbers) that can be put in such a correspondence is said to be ***countable,*** or ***denumerable.*** The act of constructing such a correspondence is the act of counting, or determining "how many" things are in the collection. The answer, a number, is called the ***cardinality*** of the collection, and the number representing this amount is a ***cardinal number.***

For example, our familiar positive integers are cardinal numbers, albeit finite ones. Three is the count of one, two, three, an intrinsic listing that exhausts any threesome. However, as Euclid first showed, our listing of the primes is never-ending and ever-reaching. Each prime patiently awaits its turn, so that a first description of the countable infinity of the primes is thus represented by a cardinal number different from any finite number. This cardinal number is usually written as $\aleph_0$, read "aleph zero." The iconography mixes the beginning of the Hebrew alphabet with the beginning of number to create a symbol for the cardinality of any infinite but countable collection. It is "beyond finite," and thus called a ***transfinite number.***

The theory of transfinite numbers was laid out by the German mathematician Georg Cantor (1845–1918). He was the first to investigate arithmetic in the transfinite realm, showing that the countable infinity represented by $\aleph_0$ was the smallest member of a new and mysterious transfinite number system. Since both the natural numbers and the prime numbers are countably infinite in extent, we see that transfinite accounting allows for the part to be equinumerous with the whole. Moreover, even the collection of all nonnegative fractions, represented by all pairs of natural numbers, and hence seemingly much larger than the natural numbers alone, is still countable, a fact demonstrated by a listing of the fractions according to the size of the numerator and denominator.* It is not until we consider collections such as the set of all possible numbers between zero and one that we begin to find transfinite numbers which exceed a countable infinity. This fact is demonstrated by the consideration of any putative listing of the decimal expansions of all such numbers. Such a list cannot include that number which differs from the first number in the list in the first decimal place, or from the second number in the second decimal place, and so on and so on.

Asymptotics: how do the primes grow?

The characterization of the primes as possessing a countable infinitude gives a quantitative description of their infinity, but what of the quality?

For example, how does the countable infinity of the primes differ from that of the odd numbers, which also can be listed one by one? The odd numbers go marching on, regularly spaced at every other positive integer, but the primes seem to have no such simple rhythm. Twenty-three is prime; the next prime does not appear until six steps later, at twenty-nine; but after that the next prime is at thirty-one, only two steps farther. The primes twenty-nine and thirty-one are an instance of ***twin primes,*** prime pairs that are but two steps apart. Another such pair are the primes 227 and 229. Twin primes have been found way out in the numerical stratosphere, with the current largest pair having more than 24,000 digits. That is to say, these two numbers would require more than ten pages of this book to write down, both are prime, and they differ only in the last place. Numerical experimentation supports the widely held but still unproved belief that there are an infinity of twin primes among the natural numbers. Such

*Such a listing would begin (omitting repetitions) like this: 0/1, 1/1, 1/2, 2/1, 1/3, 2/3, 3/1, 3/2, 1/4, 3/4, 4/3, 4/1, . . . (Notice that 2/4 is omitted because 1/2 is already included; and 4/2 is omitted because 2/1 is already included.)

brief paroxysms of primality are but one reflection of the irregular appearance of the primes. The rhythm of the primes is more a jazzy syncopation than the steady beat which describes the odd numbers. Edgar Allan Poe once wrote that "there is no exquisite beauty without some strangeness in proportion." The uneven footfall is at the heart of the allure of the primes.

Thus, while both the odd numbers and the primes are countably infinite collections of numbers, the way in which they are distributed among all natural numbers differs. So now we ask a new question: Is there a way to describe the rate at which the primes occur? Can we give a law that predicts the appearance of the primes?

Imagine the number line as an east-west sidewalk bordering a highway that extends from horizon to horizon, an infinite path marked evenly by integers, signposts spaced one yard apart. Euclid tells us that as we walk step by step, forward along the number line, no matter how far we walk we are always certain to come upon another prime. But when? Is there a roadside billboard claiming "Next prime, two integers ahead" or "Next prime, 2,000 integers ahead"?

Prospecting for primes, we march forth with a basket in hand, examining each roadside integer, leaving in place those beautiful numerical hybrids that are the composites, looking only to pick the primes—which, in the words of the number theorist Don Zagier, "grow wild like weeds." As we advance, our basket grows heavy with primes, but the farther we go, the less frequently we appear to add to our collection. This is predicted by Eratosthenes' sieve, for as we stroll along the number line, each prime we encounter foreshadows an infinity of places ahead that cannot be prime. We reach two and toss it into our basket, knowing that of the numbers ahead, every other now is ruled out: four, then six, then eight, then ten, and so on. As we march forward from two, the first nonmultiple of two we reach is just ahead at three. As three is not a multiple of any of the primes currently in our basket, three must be prime, so it is added to our collection. As before, we look forward and anticipate the successive encounters every three steps ahead of multiples of three, another evenly spaced infinity of composite numbers. Thus does each encounter with a prime initiate a perfect wave washing away the regularly spaced composites ahead.

We keep a tally of the contents of our basket and track the accumulation of the primes. One prime in the basket as we pass two, then two primes in the basket as we pass three. The count remains at two until we pass the number five, whereupon we now have three primes in the basket. The contents of the basket increase to four as we pass the number seven. Then it is not until we go another four steps and reach eleven that we'll find another prime and thereby bring our tally to five, indicative of the five prime numbers that are at most eleven.

If we knew the rate at which the primes appear, then we could also deduce the rate at which the size of the primes grows. For example, if we are told that there are ten primes that are less than thirty, but eleven primes no greater than thirty-one, then we know that the eleventh prime is equal to thirty-one.

The irregular appearance of the primes reflects the varying intervals of time over which the contents of our basket do not change. We sense a rate of accumulation that generally appears to be progressively slower and slower, although intermittently interrupted by seemingly random hiccups of surprisingly short increments. How to quantify the pace of the primes?

A simple listing of numbers gives one picture of the accumulation of primes but still comes up short as a means of revealing any structure in their growth. Such a list is like a diary that purports to record the progress of a newborn baby through only a daily measurement of height and weight; better would be a picture book, allowing us to see and compare the day-by-day changes. Similarly, even better than our list of numbers is a picture. It is better to replace our description of the primes with a picture that might help reveal the pattern of accumulation. A picture is worth a thousand words, and a good one here may be worth a thousand primes. So we make a mathematician's picture. We make a graph.

An east-west axis crossed by another axis running north-south creates the setting for a picture that will tell the tale of the growth of the primes.

Our number line heads off to the east, notched with the regularly spaced whole numbers. Measurements in the northerly direction serve to mark the number of primes in hand as we pass each natural number. So as we move eastward, the irregular appearance of the

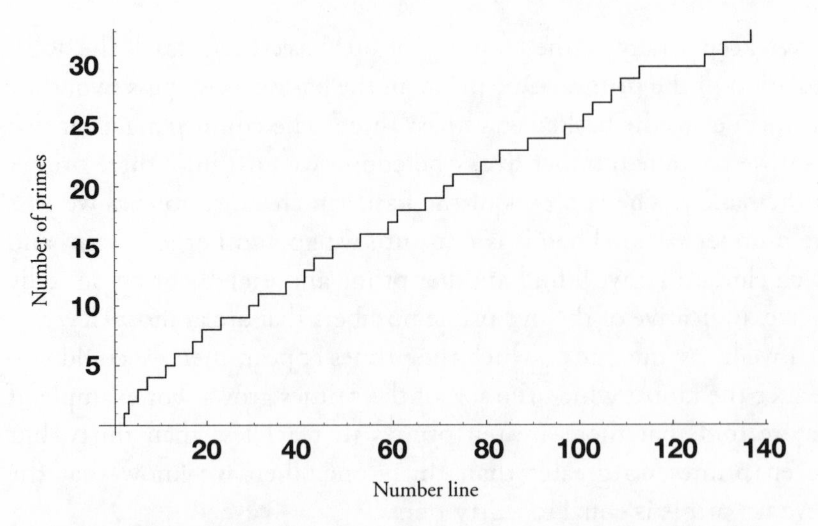

Figure 1. Punctuated accumulation of primes as we move along the number line from 1 to 137. Above any position on the east-west axis (also called the x-axis), the graph will be at a height equal to the number of primes that have been encountered up to that point.

primes gives a running count that builds a hip-hop staircase over the floor, the east-west number line. The height of the staircase at any position on the number line measures the number of primes found up to that instant. Primeless intervals, such as the short path from thirteen up to (but not including) seventeen are reflected as a plateau between these points. Here we will find a step, four units long. It rests at a height of six units, which reflects the six primes less than seventeen. But upon coming to seventeen, we find a new prime, so that our number of primes increases to seven, and so we add a new step to our staircase.

Racing up this Dr. Seuss–like staircase would prove to be a bit difficult. In a rush we might hope to climb two or even three steps at a time. Were we hurrying up the evenly spaced steps of your office building or the local football stadium, we would soon acquire a rhythm, an even stride by which we would bound to the top. But no such luck here. We start with a step of length one, which we follow by another, only to come in turn to steps of lengths two, and two again, then four, and then two, these lengths reflecting primeless stretches. Soon we slow down, seemingly unable to develop a cadence that will anticipate the broadening primal deserts.

But we should not lose heart. Perhaps we are too close to the problem, as if Mendel had attempted to deduce a theory of inheritance from two pea plants, or Mendeleev had tried to write down a table of elements from the knowledge of only carbon and oxygen, or Darwin had tried to formulate a theory of development from looking at two finches. We need to go on and step back, in order to see the forest for the trees.

Seeing the forest for the trees

Although the minute variations seem hard to predict, order does appear to emerge from chaos if we look at things on a larger scale. In taking a step back we enter the realm of ***asymptotics,*** the study of long-term or large-scale behavior, capable of revealing the patterns that materialize as we "go to infinity." Out here, on the horizon, is where asymptotics comes into play. Just as we might distance ourselves from life's minutiae in order to gain clarity, we do the same when we use asymptotics, choosing the view from afar in the hope that from this vantage point the moment-by-moment vagaries will coalesce into a clear trend.

Asymptotics is the crystal ball of mathematical analysis, predicting the shape of things to come. We study phenomena and discern the broad outlines of what lies ahead, seeing the future all at once. Armed with this foreknowledge, we can then take the time to delve into the particulars of the past, getting our hands messy with the details that are sure to give birth to a structure already anticipated. Asymptotics enables the mathematician to understand mathematical life forward, in order to live it backward.

Like the birth of our universe, the count of primes begins in chaos, a first three minutes of erratic behavior in which the initial primal materials are revealed; but then, slowly and surely, structure seems to emerge from the primordial mess. We leave the detail of that uneven staircase of the early prime landscape and zoom out and away, hurriedly snatching primes as we speed to higher and higher numbers. The effect on our picture of the growth of the primes is one of organization, like moving from a view of New York City at 1,000 feet, say from the top of the Empire State Building, to a view from a jet plane cruising at an altitude of 30,000 feet.

Buoyed by this initial success, we continue onward and outward.

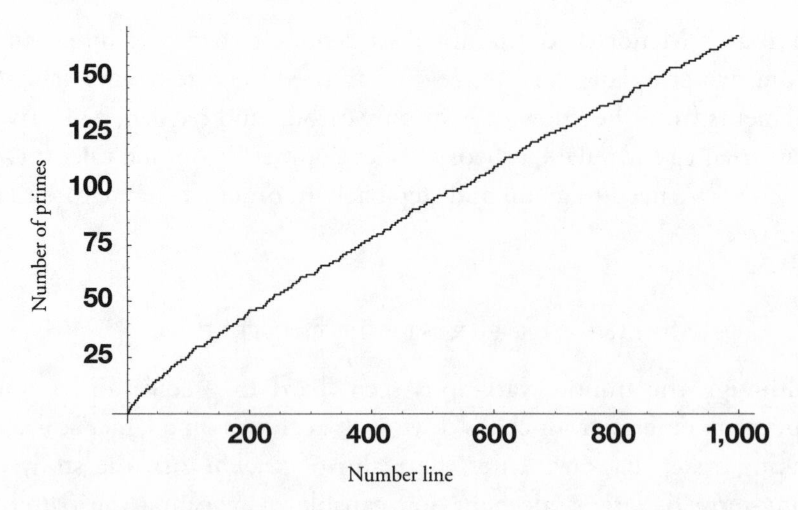

Figure 2. Accumulation of primes up to 1,000. The uneven accumulation is still apparent in the staircase-like graph.

From 30,000 feet we move to a height of tens of thousands of miles, from jet plane to space shuttle. Jagged shorelines converge to a broad, smooth, almost cartoonish outline dividing land and sea, and so too does our picture of the accumulation of the primes move from stepping-stones to a gentle curve. As we step back, the structureless becomes structured; the inscrutable now seems within our understanding. As in the distant view of a pointillist masterpiece, the details have begun to fuse into a coherent whole that we now seek to describe.

The every-otherness of the odd or even numbers means that they accumulate in a *linear* fashion, so that the picture of their count is a perfectly straight line. However, this sort of regular occurrence is not true of the primes, even from the long view. Taking hold of Eratosthenes' sieve, we might observe that generally, the farther out we go, the longer are the intervals between primes, for the farther we go, the more multiples of earlier numbers we find. This is in contrast to an infinite listing of multiples, where the distance between successive multiples is the same, no matter how far out you look.

So, in the case of the primes, the long view of their accumulation yields not a line but a subtle curve. In order to understand the asymptotics of the accumulation of the primes, we will need to bend the straight line mathematically so as to match the growth rate of the primes. In order to do this we need to find another pattern of growth,

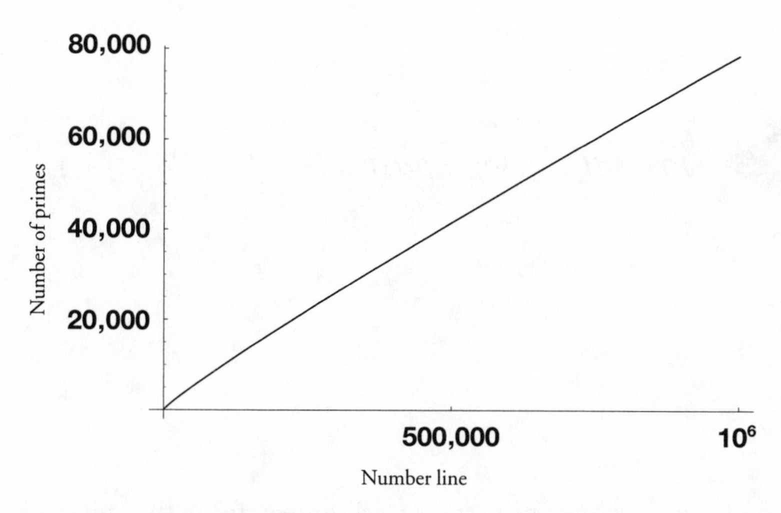

Figure 3. Accumulation of primes up to 1 million. It looks pretty close to a straight line . . . but not quite.

another endless list of numbers, and use this second list like a sequence of local directions. They will indicate that the line might be modified by subtracting a little here, or adding a little there, or multiplying by a little here, or dividing by a little there—small perturbations that accumulate to an ever so subtle deflection.

Thus, what is needed is a way in which we might slowly pull our line down from the straight and narrow; not so much that it ultimately heads to earth, but rather in such a way that it still heads out toward infinity, but at a less than speedy pace, as though we were gradually adding sand to the basket of a star-seeking hot-air balloon, to slow but not halt its journey.

So, not a line, but the most gentle of curves—that is the coarse shape of the universe of primes, but what exactly is it? What is the mathematical law that will describe this slow and less than steady growth? This is the question that would spark Riemann's interest, years after it first excited two great mathematicians of the late eighteenth century: Adrien-Marie Legendre and Carl Friedrich Gauss.

4 *Primal Cartographers*

FEW THEOREMS spring full-grown from the mind of a mathematician. Most research is a process of working out example after example in the hope of finding a common thread. Sometimes these examples take the form of reams of data that must be examined carefully in order to uncover a pattern of behavior. This is certainly the case in number theory, and especially in the asymptotic study of the accumulation of primes. The initial guesses as to the shape of its growth came from looking at great compendiums of data gathered in the form of ever longer lists of prime numbers, in the hope that the coastline, which appeared jagged from up close, might finally appear to some lucky investigator as an obvious, recognizable, smoothly growing curve.

The first cartographers of the primes were the French mathematician Adrien-Marie Legendre (1752–1833) and the German mathematician Carl Freidrich Gauss (1777–1855). Among other things, both Gauss and Legendre are known for the tools they invented to calculate the motion of heavenly bodies. Their discoveries make possible the prediction of the positions of comets, planets, asteroids, and the like from a relatively small number of observations. This penchant for and skill at data analysis was what they each brought to the problem of predicting the location of the primes in the most distant reaches of the integer universe.

Each would state a conjecture about the growth of the primes—Legendre publicly and Gauss privately. Asymptotically, the two conjectures would amount to the same thing, but as befits the precision of

Gauss, his estimate would hint at the possibility of a formula for an exact counting of the primes. Gauss made a bold and inspired guess as to the shape of the prime territory, tossing down an intellectual gauntlet that Riemann would pick up some eighty years later, when he related the shape of the primal curve to the mysterious zeta zeros.

Legendre

The first recorded prediction of the asymptotic growth of the prime counting function is attributed to Legendre. Of noble birth and broad intellect, Legendre moved quickly through the academic ranks, coming to the attention of the mathematical world in 1782 for his prize-winning research on ballistics, and the determination of "the curve described by cannonballs and bombs, taking into consideration the resistance of air."

Eventually, Legendre turned his attention to the study of phenomena of a different order of magnitude, and his inventions for doing so live on to this day. Working on a cosmic scale, astronomers still use the eponymous *Legendre functions* to help sort out the fine details in the nearly constant microwave background radiation that bathes the universe. This is the everlasting echo of the Big Bang, and in the minute fluctuations of the temperature lie keys to many mysteries of the origin and composition of the universe.

Legendre's greatest contribution to astronomy, and to applied science at large, may have been the invention of the "method of least squares," a technique invented to predict the motion of comets from a number of observations. These ideas, which he first set down in an appendix to a paper of 1806 concerning the motion of comets, provide an important general technique for finding the simplest geometric shape (such as a line, curve, or ellipse) that most closely fits a collection of data. These data might be something as ethereal as the observed positions of a heavenly body, or something as mundane as the tick-by-tick listing of a stock price or a history of corporate earnings.

The Legendre functions and the method of least squares both exemplify the interests, talents, and tools that Legendre had at hand when he turned his attention to the problem of predicting the rate of accumulation of the primes. Like a financial analyst looking to fit an earnings record, Legendre was looking to find the best curve that would fit his data on the accumulation of the primes, a curve that could predict the future growth of these numbers.

How to craft this curve? Like any good scientist, Legendre surely was driven by an aesthetic of simplicity, so he would be looking for a rule using only the basic instruments in the mathematician's curve-cutting kit. As a starting point, Legendre could see, as we saw in Chapter 3, that the primal curve is very nearly but not exactly a straight line.* Legendre observed that the primes are distributed in such a way that the graph of their accumulation bends away from linear perfection. On the primal curve, as the distance of any point from the vertical axis increases, its height above the horizontal increases, but at a slower and ever-changing rate. Legendre posited that he could effect this slight modulation by allowing the vertical displacement to equal the horizontal displacement divided by an amount that grows ever so slowly as we make our way along the number line. This is the character of something that exhibits ***logarithmic growth,*** achieved in this case by using the ***logarithm*** of the displacement, and we now turn to its explanation.

The Rhythm of the Logarithm

Logarithm. The very word can send shivers down the spine of even the most knowledgeable, but it is hardly fearsome, for as a rate of growth, called logarithmic growth, it is among the slowest of the slow. It is best understood in terms of its speedy cousin, the better known ***exponential growth,*** which derives from the exponential. Exponential growth is exemplified by the spread of a rumor told to two people, who at the next instant each tell two more people, who in turn then each tell two more people, and so on. This would be the way a tall tale might spread, lockstep, through a close-knit community: one person knows, then two more people know, then four more, then eight more, and so on. That is, first is two raised to the zero power, then it is raised

*In the Cartesian plane a line is straight only if for each point on the line, the ratio of its distance from the horizontal axis to its distance from the vertical axis is constant. This number is the slope of the line.

to the first power, then to the second power, then to the third power, and so on. In this perfectly synchronized town, the number of people just informed of the news would be described by two raised to a number equal to the number of instants elapsed.* There is nothing special about the number two. At each step, any person just informed of the rumor could have relayed it to three new people, or four or five or 100. The number of people newly informed at each step by any individual is the ***base*** of this exponential growth process; the time elapsed provides the ***exponent*** with which we can measure the spread of the rumor. Thus, if each person relates the rumor to five new people, then at the fourth instant the fourth power of five, or 625, new people have just learned the rumor.

In exponential growth, the farther we look, the faster something, such as money or information, will accumulate. The logarithm is in effect the exponential in reverse, for to grow as the logarithm means that the farther out you go, the more slowly things change.

The logarithm is ubiquitous in the quantification of perception. Its rate of growth is well-suited to the relative nature according to which change seems to be experienced. For example, consider the Richter scale that is used to measure the strength of earthquakes. The Richter scale is such that our perception of the difference between tremors rating one and two on the Richter scale is the same as our perception of the difference between five and six. Nevertheless, a tremor that rates two on the Richter scale is in fact ten times as strong as a tremor that is rated one; and a tremor rated three is ten times stronger than a tremor rated two. Similarly, a tremor rated six is ten times as strong as one rated five. This constant ratio (of ten) between successive ratings on the Richter scale reflects the fact that the Richter scale measures the logarithm of the strength of tremors—i.e., it grows logarithmically in terms of the strength.

A further implication is that six on the Richter scale marks a tremor that is 10^5 times more powerful than a tremor rated one. This in turn implies that the absolute difference between tremors rated one and two is very different from the absolute difference between five and six.

*In the standard mathematical notation, $1 = 2^0$, then $2 = 2^1$, then $4 = 2^2$, $8 = 2^3$, and so on. In general at step n in the spread of the rumor, we will have just told 2^n people.

The former is essentially the difference between the numbers 10 and 100 (i.e., 90), whereas the latter is the difference between 10,000 and 100,000 (i.e., 90,000). However, in both cases the relative difference, which is given by the ratio of the two numbers, is the same: ten. Thus, while the logarithm of these two numbers grows by only one in each case, this increase of one reflects a much smaller absolute change for the smaller numbers (90) than for the larger numbers (90,000). The bigger you are, the slower you grow.

Notice that in exponential growth (with a base of two), it took only a single time step to double the number of people just told a rumor from four to eight. By contrast, to move from four to eight on the Richter scale requires a tremor that is 10,000 (10^4) times stronger.

The decibel system also works logarithmically. Our experience of a sonic increase of two decibels feels the same on the quiet end of the volume scale as at the loud end: turning up the volume on your radio from one to two feels the same sonically as turning it from ten to eleven. However, the absolute changes (in terms of the power of the sound wave crashing against our ears) are dramatically different. Quantitatively, the ratios of the power exerted in the change from one to two and in the change from ten to eleven are the same although the numerical differences are very different. To put it briefly, logarithmic scales quantify the belief that perceived change is proportional to actual relative change, an idea that sometimes goes by the name *Weber's law.*

Just as there are many possible exponentials, each depending on a given base, there are many possible logarithms. The standard choice of the mathematician is commonly called the ***natural logarithm.*** In accordance with mathematical tradition, we will simply call it "the logarithm." The natural logarithm undoes the exponential growth of the most commonly occurring base, denoted ***e.***

Although *e* may look extraordinarily uncommon, it is ubiquitous in mathematical and physical formulas. It is a number between two and three, more precisely at just above two and seven-tenths.* Thus it is not a natural number, like one, two, or three; but it is surely of nature. It is crucial to the formula that describes the shape of a loosely

*A little more precisely, the decimal expansion of *e* starts out as $e = 2.71828\ldots$ It has most recently appeared in approximate form as the amount of money, in billions of

hanging vine or necklace. In one of its most famous incarnations *e* is shown to be equal to the return on a dollar after one year, bestowed by an infinitely precise bank that continuously compounds interest at an annual rate of 100 percent.

As the exponential races off to infinity, achieving galactic distances in very short times, it must then be that the logarithm, in an effort to undo this quick work, must quickly cut down to size even very large numbers. The twentieth power of *e* is nearly 500 million, so that the logarithm of a number near 500 million is close to twenty, and numbers near 500 million times 500 million still do not have logarithms much greater than forty.

A Near-Miss for Immortality—Sidestepping the Prime Number Theorem

On the basis of his recordings of the primes, Legendre proposed a formula for counting the number of primes under a given value. We know there are four primes between 1 and 10, and eight primes between 1 and 20, but how many primes are there between 1 and 1 million? Legendre's calculations must have shown him that, roughly, the accumulation of the primes could be explained by a relatively simple statement, which would come to be known as the ***Prime Number Theorem:***

> *The number of primes less than a given value is asymptotically that value, divided by its logarithm.*

Using this simple formula as a guess, we would estimate the number of primes less than 1 million to be about 72,000, and the number less than 1 billion to be about 48 million.

Moreover, just as exact knowledge of the growth of the way in which the primes accumulate implies exact knowledge of the way in which the primes grow, so a coarse knowledge of the former yields a coarse knowledge of the latter. The implication of our proposed suc-

---

dollars, that the company Google hoped to raise with its IPO. For a wonderful history—and there is much to know—see Eli Maor's *e: The Story of a Number* (Princeton University Press, 1994).

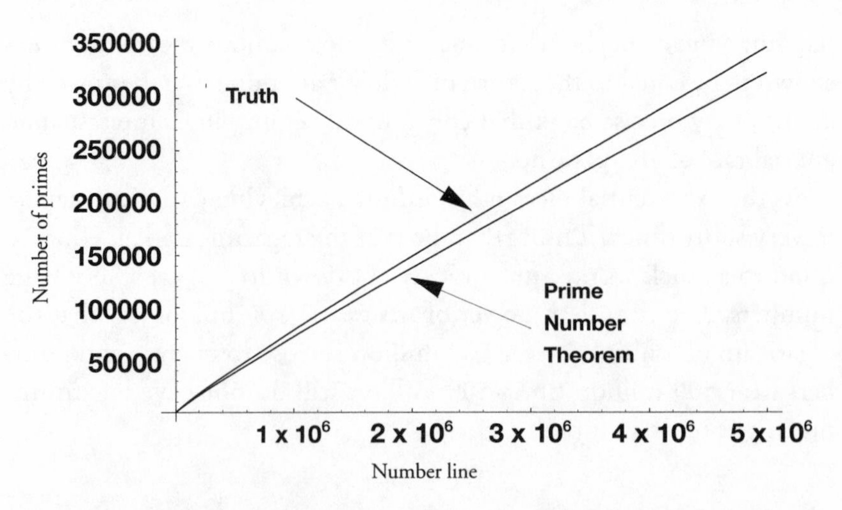

Figure 4. Graph of the exact count of the primes ("Truth"), and the graph that measures the ratio of each number to its logarithm.

cinct law of the accumulation of primes is that out on the horizon, an estimate of the size of, say, the millionth prime number is roughly 1 million times the logarithm of 1 million, or around 13,800,000. In other words, asymptotically, the size of the primes grows like the rank of the prime (i.e., where it is the list of the primes in increasing order) times the logarithm of the rank.*

Figure 4 gives a head-to-head comparison of distant views of the graphs of the exact count of the primes and the estimate due to the Prime Number Theorem. While the former does underestimate the "truth," the amount by which it misses grows so slowly as to appear negligible at farther and farther distances. In order to quantify this we can consider the ratio of the true count to the Prime Number Theorem's estimate. To say that these two are asymptotically the same is equivalent to saying that asymptotically, their ratio is close to one. Figure 5 bears out this fact.

Had Legendre stopped at the simple formula of the Prime Number Theorem, he would have henceforth been known as the first person to conjecture (but not prove!) the Prime Number Theorem, and this

*In standard mathematical language, we would write the Prime Number Theorem as follows: The number of primes less than N is approximately equal to $N/\log(N)$. Also: The Nth prime is approximately equal to $N \times \log(N)$.

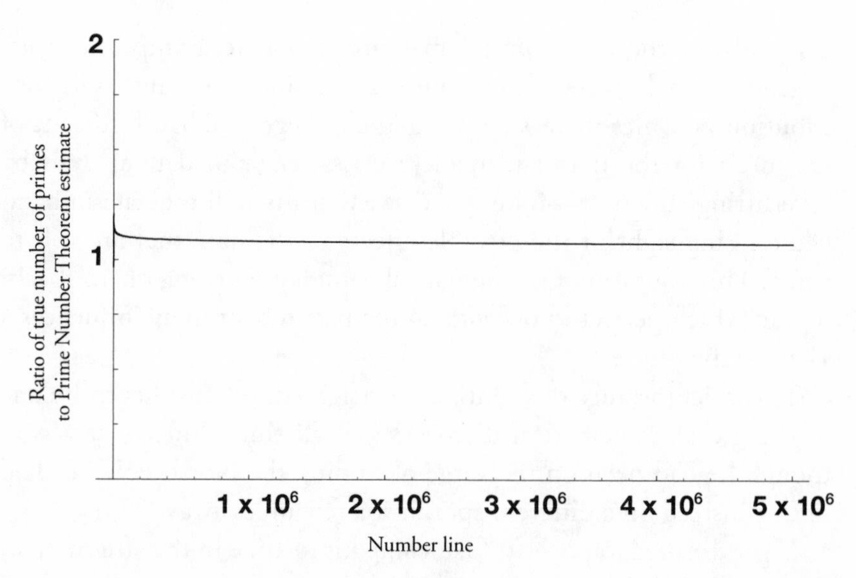

Figure 5. Ratio of the true count of the primes to the estimate given by the Prime Number Theorem. It is almost exactly one.

seminal contribution to our understanding of the primes would have more than ensured his place in the mathematical pantheon. Instead, though, he viewed this rule as a rough template for the true law measuring the count of the primes. So, like the financial analyst who, believing he sees a linear trend in earnings data, seeks to "fit a line" to those data, Legendre decided to search among a related collection of formulas for the best of the bunch. "Best" would be measured in terms of the ability of this graph to fit the data on primes that he had in hand, and so Legendre could use the same techniques that he had already developed to predict the future path of a comet streaking across the sky, now applying them to predict the course of the infinity-seeking path of the count of the primes.

Thus, in 1808, in the second edition of his text *Théorie des nombres,* Legendre published an estimate for a count of the accumulation of the primes that was close to the logarithmically inspired ratio above.*

*To be precise, Legendre assumed that the rule describing the way in which the number of primes grows would be given by a law stating that the number of primes less than a given value N is $N/[A \times \log(N) + B]$. He then found the values of A and B that best fit the data he had gathered on primes. He concluded that this best fit occurred for A equal to 1, and B equal to -1.08366, so that his recorded estimate for a formula predicting the number of primes less than a given number N was $N/[\log(N) - 1.08366]$.

Ironically, at the astronomical magnitudes where asymptotics rule, Legendre's small tweak of this simple ratio is not important, as its contribution becomes increasingly negligible. Legendre had lost sight of the forest for the trees and made a classic error in data analysis by "overfitting" the data—finding a curve that fits well the data in hand while losing sight of the possible effects of the data that are still to come. This is a form of mathematical Monday-morning quarterbacking, in which perfect knowledge of the past too strongly influences a plan for the future.

This brief instance of scientific shortsightedness cost Legendre one of the great mathematical discoveries of all time. Indeed, it was as though he had been on the verge of finding the North Pole but had settled instead on a different spot, just a few miles away.

Legendre died in poverty, first losing his fortune in the aftermath of the French Revolution, and later losing his government pension after coming out on the wrong side of a political battle. The last several years of his life seem to have been rife with fights over principle; in particular, he spent a great deal of energy battling for recognition of the priority for his research achievements, among which was his guess at the growth rate of the primes. Once again he would come up short, trumped by the same mathematician with whom he had quarreled over priority for the discovery of the method of least squares: Carl Friederich Gauss.

GAUSS: PRINCE AMONG MEN

Gauss is judged by many to have been the greatest mathematician of all time. Although he showed significant mathematical aptitude at a young age, philology and theology were his primary intellectual interests until he was well into late adolescence. It was only at the urging of his teachers that he finally decided to focus on mathematics.

Gauss's reputation comes from his contributions to the full diversity of modern mathematics, both pure and applied. Nevertheless, his nickname, "prince of mathematicians," speaks to a particular facility for the "queen of mathematics"—number theory. So it is perhaps an oedipal embarrassment that Gauss is considered the father of modern number theory. His book *Disquisitiones arithmeticae* was the primary reference in the field for decades after his death and served to introduce many a mathematician to the subject. But number theory is just the tip of the Gaussian iceberg. Even today, there is scarcely an area of mathematics that does not continue to feel his influence. He is often credited as being the first mathematician to give analytic work (i.e., the calculus) a rigorous formulation. His achievements in geometry include the discovery of non-Euclidean geometries, as well as the invention of modern ***differential geometry,*** a subject which, at Gauss's prodding, Riemann would later remake.

All this is but a brief nod to Gauss's purer work, and the problems and techniques of applied mathematics were also close to his heart. In particular, astronomy was both an inspiration for and a primary beneficiary of his work in applied mathematics. Like Legendre, Gauss was keenly interested in the problem of predicting the trajectories of heavenly bodies from finite lists of astronomical observations.

Gauss remains famous for predicting the orbit and consequent rediscovery of Ceres, a relatively small asteroid that had disappeared from sight. He took it upon himself to predict its position from its known observations and from Newton's law of gravitation. This achievement brought the twenty-four-year-old Gauss worldwide scientific acclaim in the sciences. To facilitate the pages of necessary calculations, Gauss developed a shortcut that was the forerunner of an idea now known as the *fast Fourier transform,* or FFT. The FFT is one of the most important computational tools ever invented, and one of the key concepts enabling modern digital technology. The other tool of which Gauss made a great deal of use was the method of least squares, and while he made no effort at all to publicize this grandfather of the FFT, he did take great pains to claim the method of least squares as his own, thereby initiating the long fight with Legendre for priority.

Gauss and the Primes

Gauss's failure to announce his invention of the fast Fourier transform was fairly typical of him. He is as famous for the results he didn't publish as for those he did. Gauss's strict personal criteria for publication are summed up in his often quoted remark, "One does not leave the scaffolding around the cathedral." This was his way of saying that he would permit the public to see a result only when he himself understood it so well that he could support it with a self-contained, beautifully structured proof, free of jury-rigged reasoning and flimsy argument. With such rigid requirements for publication, much of what he accomplished was relegated to his *Notebooks,* a treasure trove of research results that would have served as a publishable life's work for almost any other mathematician.

Gauss's work on the distribution of the primes must have seemed to him to be swathed in scaffolding, for the only evidence of his research on this subject comes from a letter written to a former student, the German astronomer Johann Encke (1791–1865). Encke was a well-known scientist in his own right: his name lives on in the *Encke Gap,* a region found among the rings of Saturn; and in *Encke's Comet* (first discovered by the astronomer Jean-Louis Pons in 1818), whose return Encke accurately predicted along with the periodicity of its orbit.

Encke was the director of the Berlin Observatory, and in this capacity he undoubtedly had a good deal of contact with the astronomically-interested Gauss.

Evidently, at one point Gauss and Encke had discussed the problem of the distribution of the primes. In the famous letter, dated Christmas Eve, 1849, Gauss appears to be responding to a conjecture by Encke on the asymptotic growth rate that counts the primes. It is a safe bet that Encke had in his possession a more extensive table of primes than Legendre had, and that from these he found a new estimate of the count, one which varied slightly from Legendre's overfit conjecture.*

Gauss may still have been smarting from his fight with Legendre over the priority of the discovery of the method of least squares, or perhaps he was motivated by his penchant for perfection. In any case, he claimed in his letter to Encke that fifty-eight years earlier (which would have been almost twenty years before the publication of Legendre's estimate), he had already estimated the rate of growth of

*Encke used $N/[(\log N) - 1.1513]$ to estimate the number of primes less than N, as opposed to Legendre's estimate of $N/[(\log N) - 1.08366]$.

the accumulation of the primes, and he believed his estimate to be much better than that of either Encke or Legendre.

In this letter Gauss said that as a boy of fourteen or fifteen he had taken great pleasure in counting the primes, culling them from intervals of thousands (called *chiliads*). He wrote, "As a boy I had considered the problem of how many primes there are up to a given point." From his computations he had determined that the "density of primes" near any given number "is about the reciprocal of its logarithm."

The notion of *density* used here by Gauss is akin to our understanding of population densities. In demography, *population density* refers to the average number of people living in a given unit of area (e.g., a square mile) or a particular region (e.g., a state). A density of this sort will usually vary as the region varies, in either size or location. For example, the population density of South Dakota is quite different from that of New Jersey.

Similarly, we can ask how many prime numbers "live in the neighborhood" of a particular number. Gauss's estimates imply that as we traipse along the number line with basket in hand, picking up primes, we will eventually acquire them at a rate approaching the reciprocal of the logarithm of the position that we've just passed. For example, at around 1 million (a number whose logarithm is about thirteen) we find primes at a rate of about one in every thirteen numbers; at 1 billion, we've slowed down to something like a rate of one prime for every twenty-one integers; and so on. What we now want to do is find a way to convert the ever-slowing rate at which the primes appear into a count for the number of primes that we've acquired up to any given point.

Notice that the rate at which primes occur gets slower and slower the farther out you go, but the rate at which this slowing occurs is itself very, very slow—so slow that it hardly seems to change from interval to interval. In other words, over very big intervals, the rate at which we find primes doesn't change much. This is what we'll use to try to estimate the number of primes up to a given amount.

For example, the numbers between 990,000 and 1,010,000 all have a logarithm close to thirteen, so on average, about one out of every thirteen numbers between 990,000 and 1,010,000 are prime. Thus the exact number of primes in this interval would be about one-thirteenth of the interval size, or one-thirteenth of 20,000, which is a little over 1,000. Another way of looking at this is that for the numbers between 990,000 and 1,010,000, each number "contains" about one-thirteenth, or approximately the reciprocal of its logarithm, worth of a prime. Extrapolating this argument, in order to get the number of primes between one and some given point, you might as well simply add all these reciprocals of the logarithms for all the numbers from one to the chosen stopping point.

Gauss knew how to get a very close estimate of this sum using calculus, for the idea described above is the same idea that goes into determining distance traveled solely from knowledge of the rate at which you are accumulating distance, i.e., the velocity. This is the technique of ***integration.*** By using these tools he was able to compute an estimate of the prime-counting function given by the ***logarithmic integral,*** mathematical shorthand for adding up these tinier and tinier logarithmic reciprocals.

Using the logarithmic integral as an approximation to the count of the primes, Gauss determined that for primes up to 3 million his estimate was in general better than Encke's, but worse than Legendre's. Nevertheless, given the trend in the differences, Gauss expected that eventually (i.e., asymptotically) his estimate would be much closer. Gauss was confident in his estimate, but because it had only empirical justification, he relegated it to his private personal papers. Figure 6 bears out his confidence, showing that although Legendre's estimate is more accurate initially, Gauss's is more consistent overall, and eventually is better.

The Shape of Things to Come

Gauss's formula does indeed predict that in the long run, the number of primes less than a given value is that value divided by its ratio: i.e., the exact statement of what would come to be known as the Prime Number Theorem. This is also true of Legendre's guess. However, out on the asymptotic horizon, relatively small differences are ignored. This means that two estimates can be asymptotically the same, but along the way the amount by which they differ from the true count

x	*Legendre*	*Encke*	*Gauss*
1,000,000	45	464	128
2,000,000	42	797	120
3,000,000	96	1,162	153
4,000,000	177	1,541	205
5,000,000	130	1,782	124
6,000,000	271	2,204	226
7,000,000	263	2,471	177
8,000,000	351	2,829	221
9,000,000	361	3,105	186
10,000,000	560	3,566	338

Figure 6. Amounts by which Legendre's, Encke's, and Gauss's estimates of the number of primes less than a given amount differ from the exact count. The successive rows show the differences for the number of primes less than 1 million through the number of primes less than 10 million. For example, the sixth row indicates that Legendre's estimate for the number of primes less than 6 million differs from the exact count by 271; Encke's estimate differs by 2,204; and Gauss's estimate (using the logarithmic integral) is best, differing by only 226.

can be very different. Gauss's estimate would prove to be much closer to the true count than Legendre's estimate.

The points on the upper curve in Figure 7 chart the growth of the difference between Legendre's estimate of the number of primes up to a point and the true count, and the lower curve does the same for Gauss's estimate. For example, Legendre's estimate of the number of primes less than eighty million (8×10^7) is off by about five thousand, while Gauss's estimate is off by about five hundred, and generally, Gauss's estimate is much better. Notice that Legendre's error looks to be speeding off to infinity, while Gauss's seems to stay relatively constant. In truth, it too will eventually get arbitrarily large, albeit at a much, much slower rate.

Moreover, Gauss's exact formula for estimating the primes, this logarithmic integral, holds much more information about the growth of the count of the primes than the rough guess given by the asymptotic.

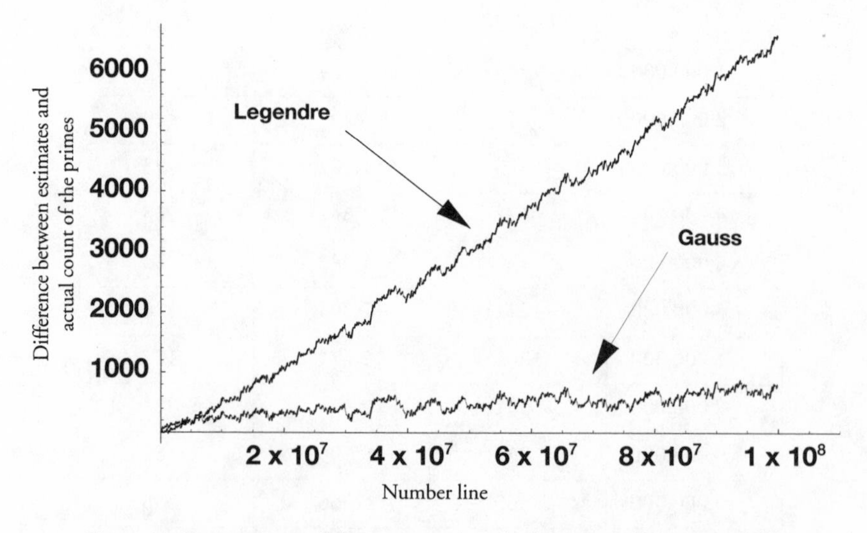

Figure 7. Comparisons of the amounts by which Legendre's estimate differs from the exact count of the primes and the amount by which Gauss's estimate differs from the actual count.

Legendre's guess shrouds the shape of the primes in a formless sackcloth, while Gauss's prediction is more a silk robe and gives the asymptotic arc a Christo-like covering that not only outlines the general behavior of the primes in the large but also gives some hint as to the true variation lying beneath.

Legendre's estimate is a dead end; Gauss's is a beginning, for embedded in Gauss's estimate of the number of primes by the logarithmic integral are the first hints of a refinement of the coarse estimate that is the Prime Number Theorem. Figure 8 takes a closer look at the difference between the true count of the primes and Gauss's estimate out to 10 million.

If in keeping with the asymptotic spirit of things we neglect the frenetic local ups and downs, a clear slowly growing curving trend is visible. This represents the first detail in the description of the growth of the count of the primes, the first tantalizing stirrings of the lively variation missed by the Prime Number Theorem. The shape underlying this first variation must itself be determined if we are to uncover the true count of the primes. Indeed, after obtaining this next level of truth, there is still more to do. A next best approximation begets the

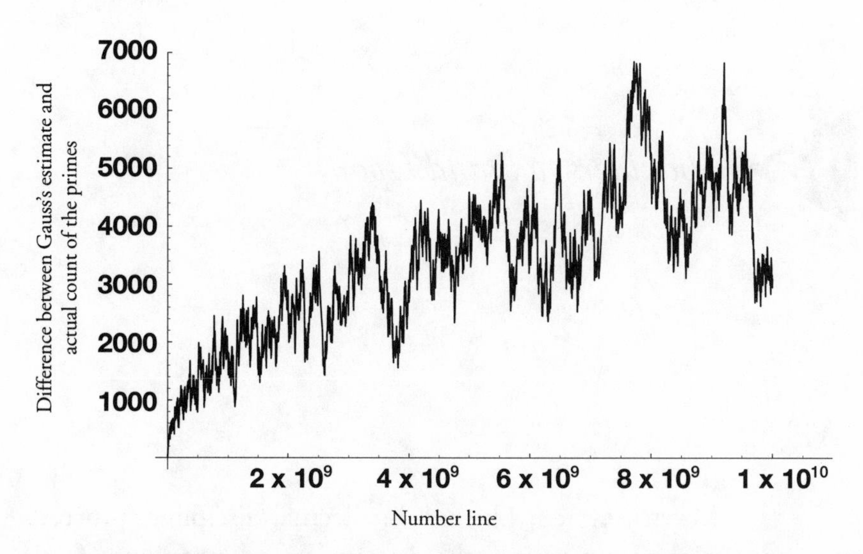

Figure 8. Absolute difference between the true count of the number of primes less than a given amount and Gauss's estimate. Riemann would set his sights on distilling the implicit underlying curved trend.

need for another and then another. Getting to the truth at the end of this process is what would excite Riemann and would be one of his great achievements, but he wouldn't do it by himself. Leonhard Euler and Lejeune Dirichlet would provide the first indications of the right direction of investigation.

5 *Shoulders to Stand Upon*

MATHEMATICS, like any intellectual discipline, proceeds by a process of accumulation, the discoveries and assertions of one generation building on those of its predecessors. The great scientist and mathematician Sir Isaac Newton once famously said, "If I have seen farther than others, it is because I have stood on the shoulders of giants." Such is the process of science, the discoveries of one generation serving as a legacy for the next. Some scientists leave tangible, hard-won treasure chests of data; others might leave a map or a cryptic deathbed remark, hinting at a land of untold promise and reward.

Many the budding scientist hears the story of the apple falling on Newton's head, jarring a connection of ideas in which Newton saw the apple's analogy in the moon moving around the earth, or the earth around the sun, each forever falling toward yet never colliding with some massive attractor. This scientific legend is a romantic version of what was surely a hard-fought intellectual assault that placed Newton atop a pyramid of astronomers, mathematicians, and physicists. For he succeeded in developing a theory whose foundations reach back to the ancient scientist-priests who first laid out the heavens according to constellations and then organized their motions in the first space-time theory: the calendar. Fast-forward to the sixteenth century, when a law of elliptical orbits discerned by Johannes Kepler from the careful and copious astronomical observations of Tycho Brahe would serve as inspiration but would ultimately be recast by Newton as a consequence of a general law of gravitation, whose derivation would require the additional invention of a new mathematical science, calculus.

Newton's is but one story in the saga of science, and it is a story which is replayed in our tale of the primes. Newton's laws predicted the motion of the planets; Riemann's law will predict the appearance of the primes. Newton had his Kepler and Brahe; Riemann had Legendre and Gauss. Newton's calculus can trace its roots as far back as the Greeks but builds directly upon that of his friend, mentor, and predecessor at Cambridge, John Barrow. In order to move beyond Gauss, Riemann would push off from the Greek-inspired work of the great Leonhard Euler, as well as that of his own friend, mentor, and academic predecessor, Gustav Dirichlet. Together they would show the way, bringing the tools of calculus to bear on the problem of the primes. They would provide the first sextants for the primal explorers.

EULER: A BRIDGE-BUILDER

Leonhard Euler (1707–1783) was a Swiss mathematician and one of the all-time greats. Prolific in every sense, over his lifetime he produced thirteen children as well as forty-eight volumes of work, much of which is still being edited and mined for hidden gems.

For Euler, like Gauss, and then Riemann, the various areas of mathematics were all of one piece—a startling range of achievement in light of what is today a highly Balkanized world of science. There is scarcely an area of mathematics that was not touched by Euler's intellect. The great mathematician Pierre-Simon Laplace is reputed to have said that the only way to learn mathematics is to read Euler, "the master of us all," a title by which Euler is still known.

Euler moved across the mathematical landscape, creating whole disciplines when necessary, building bridges between what had appeared to be isolated islands of research. The mathematics used today to model the natural world is rife with the name Euler. Scientists still study the *Euler equations* in order to understand the complexities of hydrodynamic flow, and models of the evolution of climate and phenomena like El Niño make use of *Euler-Galerkin methods.*

A mathematical Milton, Euler lived his last eighteen years in darkness, blinded by cataracts, but continuing his ceaseless work through dictation to his son, Jacob. During these last years of his life, Euler laid to rest one of the then great unsolved problems in *celestial mechanics,* the mathematical subject which studies and models the motion of the heavenly bodies. Euler was able to calculate the orbit of the moon as influenced by the gravitational pull of both the sun and the earth. This was the first instance of a solution to Newton's three-body problem (the two-body problem is effectively solved by Newton's equation for the gravitational attraction of two masses upon each other). Though blind, Euler was still able to see in his mind's eye the movement of these heavenly bodies and to make that movement visible to the rest of us in a complex mathematical poetry of equations. Awake beneath an internal night sky, Euler spun an epic tale of the moon held in thrall to the forces of gravity and so forever falling toward the earth and sun, while Milton, in his own blindness a century earlier, turned his thoughts to God and told us of the fall of man, simultaneously pulled toward paradise and hell.

With these sorts of achievements Euler succeeded in finding simple but relevant models for real-world phenomena. Relevant, but not exact, for rarely is any real part of the world exactly modeled by mathematics. One purpose of abstraction is to create a model that extracts the essence of a phenomenon while ignoring the details of the particular instantiation. Perhaps no area of mathematics more aptly epitomizes this outlook than ***topology,*** a subject whose creation is often credited to Euler.

As Euler defined it, topology—or, as it was classically called, *geometria situs* ("the geometry of position")—is concerned with those properties of position that do not take magnitude into account. Geometry requires the notions of angle and distance, or a *metric,* and in this sense is rigid. Objects are differentiated by microscopic changes. Topology is a more forgiving subject. Objects that can be deformed continuously one into another are considered one and the same. A bagel is a doughnut is a wedding band is a hula hoop. But a bagel is not a pretzel.

The beginnings of topology are most often traced to Euler's solution of a famous little puzzle, the seven bridges of Königsberg. Euler determined that it was impossible to find a walk through the city's system of bridges (part of its parks system) which would traverse each bridge exactly once. His solution of this brainteaser reveals the breadth and depth of his intellect. He strips away the real-world trappings to focus on what is important. Landmasses are shrunk to ideal points, or *vertices,* and bridges are shrunk to lines, or *edges.* The beautiful parks system of Königsberg is replaced by a ***graph.*** This graph is different from the graph of a function: it is, instead, a representation which retains only the relevant relationships of connectivity.

Euler's investigations mark the beginning not only of topology but also of ***graph theory,*** a subject which has been revivified in the modern analysis of networks. The Internet is a huge graph; wires become edges; hubs are vertices. We send e-mail to one another, flooding landlines and airwaves with a rush of bits and bytes. Fittingly, the man who laid the mathematical foundations for the study of the conduits of this electronic tsunami is also the one who first modeled the simple wash of water from one end of a pipe to the other.

These great achievements have in common an inspiration drawn from the world around us, but Euler also drew inspiration from within, and he was just as excited to investigate the intrinsic nature of mathematical objects. This is the man who wrote the book *Analysis of the Infinite,* and his other nickname, "master of the infinite," speaks to the delight with which he pursued the study of the ***infinite series,*** a cornerstone of the calculus. Moreover, it is through the infinite series that Euler connected number theory with analysis. In so doing, he provided Riemann with a first major clue for attacking the puzzle of the distribution of the primes.

Connecting the Dots: The Continuous Meets the Discrete

While mathematics does share characteristics with its scientific cousins, there are also differences. Certainly one way in which mathematics differs from the physical sciences is that mathematics produces theorems, not theories. A theory explains that which we can currently see, and makes predictions of things yet to be seen, events whose appearance will at best confirm an idea, and at worst prove it wrong. Aristotle's description of the basic constituents of matter as fire, earth, air, and water eventually falls to Lavoisier's notion of elements. This in turn gives way to the atomic theory of Dalton, which steps aside for Bohr's electron-nucleon model, which ultimately yields to Gell-

Mann's world of quarks of various strangeness and flavors. A theory of the heavenly motions is similarly evolutionary: Ptolemy gives way to Copernicus, who steps aside for Newton, who tips his hat to Einstein.

On the other hand, 3,000 years ago Euclid proved that there are an infinite number of primes, and his proof stands up today and will stand up tomorrow and forevermore. It is a fact not of a time, but of all time. Theorems are stuff for the ages, but a theory is only as good as its most recent experimental confirmation.

The eternal nature of mathematical truth is irresistible to some, a sliver of immortality which in its cartoonish extreme has grown into the portrayal of mathematicians as arrogant pedants in films like *A Beautiful Mind* and *Good Will Hunting*, or plays like Tom Stoppard's *Arcadia* and David Auburn's *Proof.* Conversely, thoughts of eternity can also be daunting. The proof rendered almost always seems so messy in comparison with the pure beauty of the perfect theorem; as a result, stories of Gauss-like perfectionists abound in mathematics. Perhaps this is the source of the aesthetics of brevity in mathematical proof. The shorter the argument, the closer it seems to the economy of thought that is the flash of insight, which is intuition.

Nevertheless, for all its trappings of eternity, a proof of a mathematical fact does not necessarily lay to rest a subject or problem. We may know something to be true, but it is not necessarily then set aside like a specimen in a museum, hidden away only to be admired or trotted out as needed. The best theorems mark not the end of a quest but the beginning, and perhaps the frequency with which a known fact is revisited is the best measure of its importance or centrality. The prefect portraits of Vermeer do not make unnecessary the canvases of Picasso or Rothko; the sensuous figures of Rodin do not obviate the forms of Brancusi or the wraithlike bronzes of Giacometti. Each great artist makes his or her own comment on the human form or the human experience in a different way, revealing a different aspect of the truth of existence. Powerful, groundbreaking work is less the actual accomplishment than the future work that it enables. It is less the sight than making possible the ability to see anew.

Proofs are themselves mathematical objects, and so a proof, if beautiful, if useful, will show the way to many more interesting results, suggesting connections between areas and new techniques to solve different problems. In many ways, mathematics is less the body of theorems than the reasoning by which they are supported. It is more means than end, more trail than mountaintop.

Euclid revealed the infinitude of the primes, but his approach lacked subtlety. Plowing straight ahead like a bulldozer, he had effectively bypassed the more delicate questions of the nature of this infinitude. Given any finite collection of primes, Euclid had shown that there was a way in which a prime larger than any of those could be uncovered. While this implies infinity, there is no part of the explicit procedure which is itself innately infinite. Two larger questions—How infinite are the primes? How fast do they grow?—are fundamentally asymptotic questions, i.e., questions about the long-term behavior of infinite processes, and questions that the infinity-phobic Greeks would not ask. In this way Euclid's proof is certainly of his time.

It wasn't until 2,000 years later, in Alpine Switzerland, that Euler found a new route to this mountaintop of number theory through the subject of ***analysis.*** Analysis, like the calculus from which it derives, is the study of the continuous and the infinitely small. Euler's great contribution to our story of the primes is as a bridge-builder, for it was Euler who first saw a way to connect the mathematics of the calculus—smooth and continuous—to that of the primes, which is jagged and discrete. He had the imagination to reverse Euclid's telescope in order to see the primes anew.

An Upside-Down Look at the Primes

Euler turned Euclid's proof of the infinitude of primes on its head. Instead of looking directly at the prime numbers, Euler considered their reciprocals—one-half, one-third, one-fifth, one-seventh, one-eleventh—and asked a different question: What happens if I start to add these fractions?

Euler realized that if he could understand how this sum accumulated, it might be possible to begin to understand the rate at which the primes appeared. For example, suppose that the primes were very, very rare, so that as we traveled far, far out along the number line, the vistas between the primes generally became increasingly enormous.

This extreme sparsity would imply that in a listing of the primes, the individual numbers soon would be huge. These huge primes produce very tiny reciprocals. Primes of a size on the order of millions have reciprocals on the order of millionths, and so on. So, if it were the case that primes were fairly rare, then a sum of prime reciprocals computed by the successive acuumulation of reciprocals of the primes taken in increasing order soon would be growing by only infinitesimal increments. Conversely, were distances between primes not tremendous, then the reciprocals would not be decreasing as rapidly, and the accumulation of their reciprocals would not tail off so quickly. In short, the rate at which the primes arise would have dramatic implications for the way in which the sum of their reciprocals grows.

This sum of prime reciprocals is an example of an ***infinite series,*** and their study is part of the foundation of the calculus, the toolbox of mathematical techniques developed by Newton to help tame nature. Newton used this sort of *infinitesimal* point of view to understand both the curves of nature and the motion of objects as a succession of better and better finite approximations, which in some ideal limit would realize the natural phenomena under investigation.

The steady accumulation of number that is this infinite sum of prime reciprocals can once again be visualized as a walk along a number line. Each reciprocal of a prime may be interpreted as a directive to walk forward by some fixed amount. First walk half a mile, then another third of a mile, then another fifth of a mile, and so on.

What can we make of such a journey? Maybe it is like the steady march described by orders to travel one mile, then another mile, and then another mile, and continuing this forever, a journey in which we tromp off to the horizon well aware that with patience we will one day reach, and then pass, any goal which we set for ourselves. On the other hand, perhaps it is instead like the seemingly infinite journey to the other side of the room mapped out as a directive to first walk half the way, then from this new position half the way again, and on and on, resulting in an infinite incremental journey that eventually comes as close as we like to the other side while never reaching the destination; our goal is the gold ring that is forever in sight, but forever out of reach. In this way the world of the infinite can be rife with frustration, confusion, and paradox, which the ancient Greeks found so disturbing that they forbade its study and discussion. Their "abhorrence of the infinite" is traced in part to thought experiments like this never-ending half and then half again trip across the room, which the philosopher Zeno used to construct paradoxes capable of "proving" the impossibility of motion through a misunderstanding of the infinite.

Rather than a source of confusion, such considerations were a fount of pure joy for Euler. He delighted in working with the infinite, and often succeeded in making mathematics out of its mysteries. Zeno's half the way, then half again walk is an example of an infinite series that *converges,* and the corresponding sum of one-half to which is added one-fourth, and then one-eighth, and so on is given the sum (or "limit") of one. By contrast, the steady unbounded march is an example of a *divergent* series.

Euler bushwhacked out on this mathematical frontier. Disregarding the rules of polite mathematical behavior, he acted somewhat as a mathematical outlaw, often making intuitive leaps without sweating over the details that would consume his followers. Euler paid little attention to the rigor and cries for certainty that would be the hallmark of the great German mathematical analysts, and in this borderland between truth and proof, he managed to stake out a new homeland for the world of mathematics. This would be a place in which number theory would flower, seeded by Euler's study of the distribution of the primes through the analysis of a particularly fascinating infinite series, the ***harmonic series,*** which would ultimately provide the raw material with which Riemann fashioned the key to understanding the primes.

The Harmonic Series

Euler's analysis of the infinite accumulation of prime reciprocals started with a mathematical fact that had been known to mathematicians for hundreds of years: if a person tried to add up all the reciprocals of the integers, adding one-half to one, then one-third to this, and then one-fourth to that, and so on, then (just as in the case of an infinite series modeled on the steady mile-upon-mile walk to the horizon), this sum of reciprocals would ultimately though very slowly surpass any preassigned amount.

This summation of all reciprocals is called the harmonic series, honoring its relation to the Pythagoreans' numerical-musical mysticism. The Pythagoreans discovered that two strings which are commensurable in length (meaning that the lengths are simultaneously whole multiples of a single common length) will sound harmonious when struck simultaneously. For example, one string might be twice as long as the template, and another string three times as long. These lengths are commensurable and, in particular, are in a ratio of two to three. Continuing along this path of harmonious commensurability, we build a many-stringed guitar by snipping off as many jointly harmonious lengths of string as we please: start with one string, then add another whose length is one-half that, then add another whose length is one-third the first, and so on. At some point we stop adding strings, and the chord we get by simultaneously strumming all those strings will be music to our ears.

The harmonic series describes how much string would be needed if we wanted to build an imaginary guitar using strings of all possible reciprocal lengths. To see that indeed there isn't enough string in the world, consider the successive amounts of string needed for the first two strings, then the next two, and then the next four. Counting lengths off in feet, our first string will require one foot and our next string another half foot. The next two strings are one-third and one-fourth of a foot respectively. Now we make a simple observation: since one-third is bigger than one-fourth, the addition of these two strings requires that we use at least an additional half-foot of string.

We now quickly count off the amount required for the next four strings. It is exactly the sum of one-fifth, one-sixth, one-seventh, and one-eighth foot. Since each of these four lengths is greater than or equal to one-eighth, their total length is greater than or equal to that of four pieces of string each of which is an eighth of a foot long, which would be another half a foot. So, in order to make the next four strings we need more than another half-foot of string.

At the next step we will look at the total length needed for the next eight strings; after that, the following sixteen strings; and so on, at each point doubling the number of strings we considered at the previous step. If we keep going like this, we find that we continue to accumulate string in chunks that exceed one half-foot of string.* This means that our Platonic guitar requires an infinite amount of string, or equivalently, that the harmonic series grows to infinity.

A Harmonious Proof for Euclid

Where do the primes enter into the harmonic series? How can it be that the chord played from the harmonic series has embedded in it a music of the primes? Well, just as all integers can be created by multiplying together all possible finite collections of primes, so too the harmonic series, which is the orderly accumulation of the reciprocals of each and every integer, can be expressed not just as a sum but also as a product—a product requiring an infinity of factors, one for every prime. This is the famous ***Euler factorization*** of the harmonic series.

Each factor comprises the entire contribution to the harmonic series corresponding to a single prime number, and is itself another infinite series that is for each prime much like Zeno's half and then half again walk. This walk of one foot, then half that, then half that again (for another quarter), and half that once more (for one-eighth), and so on, which comes as close as we like to advancing a whole two feet, makes sense of the statement that the infinite sum of the reciprocal powers of two is two:

$$1 + 1/2 + 1/4 + 1/8 + \ldots = 2$$

Similarly, were we instead to walk one foot, then one-third that, then one-third that again (for one-ninth of a foot), and one-third that again (for one twenty-seventh of a foot), and so on, we would approach as close as we like a distance of one and one-half feet. For one last example, the same procedure using a factor of one-fifth at each step brings us ever and ever closer to a point five-fourths of a foot away. In short, we say that the sum of all inverse powers of three is one

*What we are edging toward here is that at step N of this process, the length of each of the next 2^N strings is at least $1/2^{N+1}$, so that their total length is at least the product of these two numbers, or 1/2.

and one-half, and the sum of all inverse powers of five is one and one-quarter.

We do this for every prime, considering walks where at each step we advance a fraction of the distance that we advanced just one step before, with the fraction equal to the reciprocal of the prime. In each case (as for two, three, or five) there is a point which we approach as close as we like, but never surpass, and this is the sum (or infinite series) of the reciprocal prime powers. These are the factors of the Euler factorization, one piece for each prime, and when they are all multiplied together their product yields the song that is the harmonic series.

This is a song of infinity, for, as we've already seen, the harmonic series grows without bound. So were there but a finite number of primes, then the product of their terms (each of which is finite) in the Euler factorization would necessarily be finite. For example, were five the biggest prime number, then, using Euler's idea, it would be possible to express the harmonic series as the product of the three numbers derived above: two (for the sum of the reciprocal powers of two), three-halves (for the sum of the reciprocal powers of three), and five-fourths (for the sum of the reciprocal powers of five). Their product is thirty eighths. This is a fact that is terribly at odds with the divergence to infinity of the harmonic series. Since mathematics cannot tolerate two contradictory facts, there must be an infinite number of prime terms. Thus, once again are we assured of the infinity of the primes. Euler has found a new road to a truth first uncovered by Euclid.

Building a Series of Primes Out of Logs

Perhaps this whispered hint of primal melody in the harmonic series would have been enough to lure mathematicians to investigate it, for a new proof of a known important fact is always of interest. But Euler went even further. In doing so, he showed his awareness of the deep connections between the rate at which the primes grow and the properties of the harmonic series and its generalizations, making a great leap in an intellectual and mathematical journey that would find its culmination in the work of Riemann.

Euler rolled up his sleeves and dug into his new proof of the infinitude of the primes. Using his mastery of the manipulations of the infinite, he was able to use the Euler factorization to extract from the harmonic series only those terms (or string lengths) of prime reciprocals, such as one-half, one-third, one-fifth, and one-seventh, thereby considering only that Platonic guitar whose strings have lengths corresponding to prime reciprocals.

The relative sparsity of primes as we travel along the outer reaches of the number line might lead us to believe that as we add only these reciprocals (adding one-half to one-third, and then one-fifth, one-seventh . . .) the numbers we are adding get small so quickly that, as in the case of Zeno's reluctant room-crosser, these sums all remain forever before a fixed bound. But no. Instead Euler discovered that like the harmonic series, the sum of the reciprocals of the primes also passes any preassigned point. But the rate at which we get to these signposts, or accumulate our guitar strings, is so slow as to require the patience of a yogi.

The harmonic series is already slow enough, growing roughly at the rate of the natural logarithm: the sum of its first ten terms is equal to roughly the natural logarithm of ten, which is about three; and the sum of its first 1 million terms is roughly equal to eighteen. But even slower than slow, a plan of accumulation described by adding prime reciprocals grows (and we use the word lightly) at a rate that is like the logarithm of the logarithm. Even after 1 million terms, we have accumulated only the logarithm of the logarithm of 1 million, or the logarithm of eighteen, which is a little bit more than four! This is a growth rate whose progress is like that of the snail's snail, or the slug's slug, a rate that is slower than slow.

One Series among Many

Through his use of the harmonic series, Euler built the first bridge that would connect the discrete world of number to the continuous world of calculus. This is a first analytic understanding of the way in which the primes are distributed. The growth of the accumulation of their reciprocals must tell us something about the rate at which they appear at the far reaches of the number line.

But Euler's contribution goes beyond being the first one to put a stopwatch on the primes. Euler, Master of the Infinite, saw the harmonic series not as an isolated piece of infinite amusement, but rather as one possibility among a vast and never-ending continuum of infinite series. He started by considering the sum of the reciprocals of the natural numbers, which is the harmonic series, but why not other related accumulations too? What about the squares of the reciprocals? The cubes? Why not consider the sum of the reciprocals of any power?*

It turns out that the behavior of the prime reciprocals is different from all these other sums. For example, suppose that only the reciprocals of square numbers are added, a walk across the room for Zeno's persistent student that demands first a step of one foot, followed by one-fourth (i.e., one over two squared) of a foot, and then another ninth, and so on, each time moving ahead by an amount equal to the reciprocal of the next squared number.

In fact, this sort of walk does stay within the realm of the finite, never surpassing (and coming as close as we like) to a total distance of one-sixth π (pi) squared, a number that is slightly larger than one and one-half. Similarly (and mysteriously), walks that use other even powers (such as directing that at each step you move forward according to the reciprocals of fourth powers, or sixth powers) yield other fractions of powers of π.

The implication is quite surprising: since the sum of the reciprocals of the primes passes any point, but the sum of the reciprocals of any fixed power is finite, then it must be the case that the primes are more densely distributed than any power. This is one of the first recorded facts regarding the density of the primes.

By considering the sum of the reciprocals of the squares, and the cubes, and so on, Euler showed that in fact the harmonic series was but one species in a newly discovered genus of infinite series, that the

*In other words, Euler looked at series like the sum of reciprocal squares,

$$1 + 1/2^2 + 1/3^2 + 1/4^2 + 1/5^2 + \ldots$$

the sum of reciprocal cubes,

$$1 + 1/2^3 + 1/3^3 + 1/4^3 + 1/5^3 + \ldots$$

and more generally, the sum of reciprocals raised to any natural-number power.

harmonic series was but one point of an infinite family of infinite series, one for each natural number. Riemann would have first seen this reading Euler's *Analysis of the Infinite* as a schoolboy. It would ultimately prove to be the starting point of his own reinvention of the study of primes, but it was Riemann's true academic mentor, Gustav Dirichlet (1805–1859) who would be the first to show that Euler's bridge between analysis and number theory was more than a footpath.

DIRICHLET: FIRST ONE OVER THE BRIDGE . . .

By all accounts Dirichlet was an amiable man, a child of the middle class, the son of an educated woman and the local postmaster. A precocious youngster, he graduated from secondary school at the age of sixteen and left Germany to study mathematics in France, then the center of mathematical research. Dirichlet had the repeated good fortune of finding one generous mentor after another. He made steady progress in his work and was soon able to return home, where his successes in France eased his progress up the codified stages of the German academic ladder. He assumed a highly coveted professorship in Berlin in 1831, the same year in which he married Rebecca Mendelssohn, a sister of the composer Felix Mendelssohn.

Dirichlet was able to revitalize what had become a rather moribund mathematical environment in Berlin, his intelligence and affability serving as a magnet for the mathematical community at large. After Gauss died, Dirichlet assumed the professorship of mathematics at Göttingen and remained there until his own death.

Periodic Tables for the Primes

Dirichlet enters our story as the first one to pick up on Euler's use of the infinite series in the study of primes. Euler saw the harmonic series as one of a family of possible infinite series, one for each natural number, corresponding to a consideration of the sums of the reciprocals of the squares, cubes, fourth powers, and so on. Dirichlet took this powerful idea one great leap further, realizing that this discrete collection of infinite series built from the harmonic series could be filled out to a continuum's worth of series. He did this in order to lay to rest another of the giant open problems of number theory, the problem of finding "primes in arithmetic progressions." He was thus able to begin to classify and understand the diversity of the primes.

An ***arithmetic progression*** is a list of integers that we arrive at through walks along the number line of regular step-size. The simple walk of step-size one starting at one yields a list of all the positive integers, and thus a list in which an infinity of primes occur (since all of them occur there). Should we start at two and then bound along by steps of length two, we find ourselves jumping from even number to even number, thereby obtaining the entire list of even numbers as an arithmetic progression of step-size two, starting at two. If we started at one, a step-size of two would take us from odd number to odd number, thereby producing the odd numbers as an arithmetic progression.

With respect to the primes we notice a dramatic distinction between these two arithmetic progressions. Outside the starting point of two, the list of evens contains no primes, whereas the list of odds contains all primes other than two, and in particular contains an infinite number of primes.

Now we try a new game, instead choosing to hop along the integers taking three steps at a time. If we start at one, we move on to four, then seven, and so on. Starting at two we leap to five, then eight, and so on; and finally, starting at three we go to six, then nine, and so forth. The last of these arithmetic progressions gives all the multiples of three:

$$1, 4, 7, 10, \ldots$$

$$2, 5, 8, 11, \ldots$$

$$3, 6, 9, 12, \ldots$$

Euclid had asked if the list of primes went on forever, i.e., if among the list of natural numbers there were primes as large as we would desire. We can ask the same question of our arithmetic progressions: in any arithmetic progression, are there primes as large as we desire?

To begin, we notice that of our three progressions with step-size three, the last, which comprises only multiples of three, can contain no primes other than the prime three. More generally, we see that we must restrict our attention to those progressions in which the step-size and the starting point have no factors in common.

Euler was the first to wonder publicly about the possibility of an infinity of primes in an arithmetic progression (outside the caveat mentioned above). He believed it to be possible but never proved that it was. Legendre claimed to have proved it on two separate occasions, but both times his proof turned out to be incomplete.

It was Dirichlet who finally showed that in such regularly spaced romps through the integers an infinity of primes would be encountered. In order to do this Dirichlet looked to Euler. Dirichlet wanted to mimic the way in which Euler had proved that there were an infinity of primes. Euler had accomplished this through the use of the Euler factorization, showing that the infinite nature of the harmonic series necessitated that it be made of an infinite number of factors, one for each prime. Dirichlet wanted to do the same, but this time to make an infinite series whose factorization used only primes that occurred in a given arithmetic progression. The collection of infinite series dreamed up by Dirichlet (consisting of one series for each arithmetic progression) to arrive at his results has come to be known as the ***Dirichlet L-series.*** Whereas in his other work on the primes Euler had only seen fit to use natural number exponents in his variations on the harmonic series, Dirichlet saw that he could study the arithmetic progressions by filling in the spaces on the number line, allowing himself the possibility of using sums of integer reciprocals raised to a power that could be any real number greater than one.* With this he was able to prove that an infinite number of primes appear in any arithmetic progression, except for those in which the step-size and starting place share a common factor.

Postscript

Euler's new proof of the infinity of the primes and Dirichlet's further exploitation of Euler's techniques to find deeper structure within the

*In other words, Dirichlet allowed himself to consider infinite series that looked like

$$1 + 1/2^s + 1/3^s + 1/4^s + 1/5^s + \dots$$

where s could be any real number greater than one.

prime landscape revealed the power of analytic techniques in understanding the primes. Along the way, a torch illuminating the primes passed from one occupant of the Göttingen professorship in mathematics to another. The next handoff would prove to be the decisive one. Riemann was waiting.

6 *Riemann and His "Very Likely" Hypothesis*

It is 1859, a landmark year in the history of science. Charles Darwin publishes *The Origin of Species* and thereby publicly announces a scientific theory capable of characterizing the growth and development of the natural world. His theory of evolution arrives on the scene as an intellectual thunderbolt, the culmination of a steady accretion of data and data analysis, using the notions of fitness and selection to explain the minute variations as well as grand temporal trends recorded in this naturalist's detailed notebooks.

This same year, a bit east of the other side of the English Channel, the earthshaking achievement of a diffident British biologist finds its mathematical counterpart in a brief paper, "On the Number of Primes Less Than a Given Magnitude," presented to the Berlin Academy by a German mathematician, as shy and softspoken as Darwin. The accompanying lecture, delivered in honor of his election to the rank of "corresponding member" of this august body, marks the revelation of a new idea for explicating the growth patterns implicit in the tables of primes and composites, the flora and fauna of the Platonic world of the natural numbers.

This "Darwin of number theory" surprises the leading scientists of the day by announcing the discovery of a beautiful formula, one that can illuminate the irregular hiccups which mark the punctuated evolution of the primes and can also illuminate their fusion into a broad overarching picture summarized by the conjectured, but still unproved, "Prime Number Theorem." The discoverer of this formula is Bernhard Riemann, and his paper contains what is the now famous ***Riemann hypothesis.***

Like its Darwinian döppelgänger, Riemann's paper is an epochal scientific achievement. With an outlook akin to that of his teacher Gauss, Riemann sought a description of the distribution of the primes capable of more than a coarse-scale replication of their accumulation. But like any aspiring scientist, Riemann surely wanted to do more than chisel one more detail into the cathedral of the past achievements of others. His goal was no less than an illumination of the whole truth, from the large to the small, from the far-off asymptotic grand sweeping curve to the zigzag nooks and crannies of close detail. His discovery of an exact formula, with the concomitant implications of his hypothesis is, like Darwin's theory of evolution, one of the great scientific signposts, a place from which we measure a before and an after in the history of ideas.

The road to the Riemann hypothesis

Unfortunately, we have no written record detailing the thought process that guided Riemann during his work on primes. We have in hand some of his notes, as well as his paper, but while not every mathematician suffers Gauss's need for cathedral-like perfection, most mathematics papers are largely "scaffolding-free," often stripped of any signs of intuition or thought processes. Striving more for clarity and leanness of argument than narrative, they are mostly assertion and proof and often frustratingly free of motivation.

Riemann's diary (if he had kept one) might have tracked the day by day and week by week wanderings of his intellectual journey, providing a road map of his frustrations and reasoning, his disappointments and happy surprises. Instead, although we have the fruits and some signs of the labor, we have relatively little insight into the process. We admire the masterful, inspired sculpture; we feel the heft of the chisels and mallets; but the muse eludes us. There is often an element of luck, a serendipitous collision of the right problem and the right problem solver. Surely there is a lot of hard work, for that is the way things usually go in mathematics: perspiration almost always trumps inspiration. In a search for a deeper understanding of Riemann's work, the number theorist Carl Ludwig Siegel (1886–1981) made it his business to wade through the reams of paper containing Riemann's posthumous notes (the famous *Nachlass,* or estate). Legend has it that upon receiving one package of notes and sifting through page upon page of scribblings and scratchings, overwritten formulas and calculations, crossings-out and erasures, Siegel turned to the mathematician G. H. Hardy and joked, "So here is Riemann's great insight." Ultimately, what we are left with is the product of a quiet unconscious formation of a web of surprising connections among raw materials of scholarship acquired over a lifetime.

The collection of Riemann's mathematical output is almost as terse as his manner—and his life, which was cut short by tuberculosis when he was just shy of forty. His published writings fill only a slender volume, two-thirds of which appeared posthumously. He lived to see only nine of his papers in print, of which just five were devoted to pure mathematics. Nevertheless, the ratio of quality of work to quantity is extraordinary. The progress of modern mathematics and physics was forever altered by these relatively few papers, among which was the jewel that led to his reinvention of the study of the distribution of the prime numbers. This would be a true but terse mathematical tour de force, displaying an economy of thought well suited to a man of boundless imagination but with no time to waste. In its brevity and unfinished form, it is like a metaphor for Riemann's own unfinished life.

Beginnings

Riemann's lecture at the Berlin Academy was one shining moment for him in a year replete with personal achievements. At the age of thirty-one he had just been named professor of mathematics at the University of Göttingen, thus following in the footsteps of Dirichlet and Gauss. This pedigree reflects the professorship's status as one of the most prestigious of the few and highly sought-after senior mathematics posts in Germany. Riemann's appointment must have been as much a source of relief as of pride. For years he had been the sole financial support for a large extended family, a situation which had stretched thin his meager academic salary. Now Riemann would have the financial and professional security necessary to allow him to turn full attention to his work.

Riemann was a man of humble beginnings, the second of six children (two boys and four girls) born to a Lutheran minister and the educated daughter of a court councillor, then living in Germany's Hanover region. His early studies were directed toward following in his father's footsteps, but an innate intellectual affinity and talent for the sciences (as well as a shyness dramatically at odds with the duties of a minister) soon overcame filial devotion. With his father's approval, Riemann took up a formal study of mathematics and physics.

Riemann cut his mathematical teeth on the classics. Legendre's *Theory of Numbers,* the massive two-volume text containing Legendre's asymptotic estimate of the accumulation of the primes, served as an introduction to number theory, and is most likely the original source of Riemann's interest in the distribution of the primes. Legend has it that he devoured Legendre's treatise in just one week, returning it to the surprised lender with a word of thanks and the offhand remark, "I know it by heart."

Riemann's own work on the distribution of the primes would knit this adolescent mathematical inspiration with the tools of the calculus, acquired from his other early mathematical influence, Euler's *Introduction to Infinitesimal Analysis,* one of the first calculus texts. Although this work was dated in approach by the time Riemann read it, the many of inspired and inspiring manipulations of beautiful, infinite formulas (including the wondrous harmonic series and many of its relatives) would serve him well in his later investigations.

Riemann's graduation from a *Gymnasium* (secondary school) was followed by enrollment at Göttingen University, where he hoped to study with Gauss. Unfortunately, the quiet Riemann and the arrogant, often acerbic Gauss must have been a poor match in temperament. After one year Riemann took leave to continue his studies in Berlin, at that time a friendlier and livelier place, recently reinvigorated by the leadership of the avuncular Dirichlet. Two years later, more mature and confident, Riemann returned to Göttingen, and in 1851 he finished his doctoral dissertation under Gauss and began his professional mathematical career.

A Reinvention of Space

At the time of the lecture in Berlin, Riemann was perhaps best known for his work in geometry, the roots of which can be found in his paper

"On the Hypotheses That Lie at the Foundation of Geometry." Like "On the Number of Primes . . ." this is a brief work—only nine pages—but it bubbles with the primordial material of an entire intellectual world. It is more sketch than finished work, more discourse than disquisition, written pithily in a tone and style appropriate to a man who perhaps feels that he has not the time to say it all precisely, and so chooses to say only the most important things in as rich a language as possible.

The genesis of "On the Hypotheses . . ." is one of those happy intellectual accidents that seem to dot the landscape of scientific achievement. Riemann wrote this paper in order to complete his "Habilitation," a piece of postdoctoral research required for advancement through the torturous trellis that was, and still is, German academe. As was the custom, Riemann submitted to his supervising committee a list of three possible topics, the last of which was an investigation of the "foundations of geometry." Traditionally, an examinee would be asked to pursue his or her first suggestion. However, as fortune would have it, the chair of the committee was Gauss, whose interest in the possibilities of geometries other than Euclidean geometry (in secret, he had already discovered the existence of non-Euclidean geometries) probably motivated his unconventional choice of this last topic for Riemann's research subject.

Riemann took up the challenge and produced a paper containing all the basic ideas of what is now known as ***Riemannian geometry.*** The classic example of a Riemannian geometry is the geometry describing the surface of a perfect sphere. Its apparently flat "local" neighborhoods (think of how the surface of a beach ball would appear to a gnat, or how the surface of the earth appears to a human being) are stitched together in such a way as to produce a domain in which even our familiar triangles acquire puzzling properties. The sphere's constant ***positive curvature*** implies a land in which the angles of a triangle sum to an amount greater than the Euclidean or "flat" paradigm of 180 degrees. For example, consider the three-sided region obtained by choosing two distinct points on the equator and traveling due north from each of them, thereby creating two lines that meet at the North Pole, and then closing the spherical triangle by joining the original equatorial points by a line hugging the equator. The lines due north form angles of 90 degrees with the equator, so that whatever the angle they form upon their intersection at the North Pole (somewhere between zero and 180 degrees), the measures of the angles of this triangle sum to something greater than 180 degrees. Furthermore, since any point on the sphere's surface can be described using two numbers, or *coordinates* (e.g., latitude and longitude), this *Riemannian manifold* is *two-dimensional.*

A sphere is the simplest sort of Riemannian space, providing the prototype of a space of *constant curvature,* but because of this constancy it is a geometry that hardly shows off the power of Riemann's invention. For Riemannian, not Euclidean, geometry is the measure of the world around us. The rigidity of Euclidean geometry is well suited for perfect lines, angles, and circles, but Riemannian geometry quantifies the local variation of the real world. Riemannian geometry makes sense of a topography that includes craggy peaks and smooth dales, rolling hills and flat deserts.

But the true power of Riemannian geometry lies in its ability to describe worlds beyond sight, spaces that shimmer at the edge of our imagination. The great geometers seem to have access to a realm beyond the senses, but just within the reach of the intellect. These thinkers are Homeric in their invention of a language able to illuminate worlds to which all must be blind. Riemannian geometry is one such great mathematical and artistic tool. By enlarging the notion of coordinatization from the two-dimensional east-west, north-south Cartesian plane to an arbitrary number of dimensions, the language and tools of Riemannian geometry mediate the familiar intuitions of two- and three-dimensional phenomena to guide investigations in spaces that are beyond direct sensory experience. This geometry brings Euclid down to earth, while simultaneously pushing Descartes to the stars.

In particular, Riemannian geometry has proved able to describe the phenomena of a relativistic world and the geometry of the space-time continuum, thereby anticipating and enabling the discoveries of physicists like Einstein and Stephen Hawking. It gives mathematical life and precision to the infinite and infinitesimal, enabling a description of a curved universe of gravitational mountains and valleys, pockmarked with black holes and looped in knots of stringlike tendrils of energy.

Riemann's sole contribution to the theory of prime numbers is in essence a creation of a Riemannian geometry of the landscape of primes. He would find a way to express the count of the primes that provides the detail of their one-by-one accumulation without missing the overall growth, a way of showing forest and trees all at once. Thus Riemann pushed ahead of Gauss and laid the groundwork for a rigorous justification of Gauss's conjectured Prime Number Theorem.

The range of tools and techniques that Riemann would bring to bear on this problem make it seem as if he had been preparing all his life for this moment; his work on the primes seems to be a nexus for a lifetime of mathematical experience. In taking on the challenge of understanding the distribution of the primes, he would pick up where his thesis adviser Gauss had left off. Riemann reached back to the writings and conjectures of Legendre that had first piqued his adolescent interest in the primes and made use of a mature application of the formulas and manipulations found in his first calculus book, written by Euler. In so doing he would extend the work of his friend and mentor Dirichlet. These advances in the study of ***complex numbers, complex analysis,*** and ***Fourier analysis*** are crucial to the Riemann hypothesis, as well as to the modern study of the phenomena of sounds and signals. We now turn to them.

A complex dissertation

Riemann's doctoral dissertation marked the beginning of his career as a professional mathematician. Gauss was the chair of Riemann's doctoral committee, and Riemann's work was sufficiently important and novel to raise even the dour Gauss's eyebrow. In signing off on Riemann's degree, Gauss remarked that his dissertation "offers convincing evidence . . . of a creative, active, truly mathematical mind, and of a gloriously fertile originality."

With this work Riemann effectively announced himself as a founder, along with the French mathematician Augustin-Louis Cauchy (1789–1857), of ***complex analysis:*** the extension of the techniques of calculus to the realm of ***complex numbers.*** This discipline is responsible for some of the most important mathematical tools in physics and engineering. Riemann's paradigm-shifting insight was the realization that it is applicable to the study of the distribution of prime numbers.

Complex Numbers

Complex numbers are, in some way, the final destination of a logical evolution of the notion of number. As we have already seen, the concept of number has its origins in what Kronecker called the God-given natural numbers. As descriptors of discrete sensory experience, the natural numbers are but a small intellectual step away from the consideration of their ratios. These ratios, numbers like one-half or two-thirds, make up the ***rational numbers,*** whose visual and tangible realization in terms of Euclid's commensurable lengths retains the familiarity and immediacy of their "natural" progenitors.

Following these rational developments things begin to get more complicated—perhaps crazy, but irrational at the very least. The Greeks discovered that there are naturally occurring numbers which cannot be expressed as the ratio of two natural numbers. Make a square, and consider the ratio of the length of a diagonal to any of the (equal) sides. It turns out that this number, which is the square root of two, cannot be represented as the ratio of two natural numbers. We have found in nature our first ***irrational number.***

Initially, the Greeks must have been troubled by the existence of the irrationals. These numbers may be ever better approximated by ratios, yet are unrealizable as such, and so embody the notion of the infinite—thus they would have been anathema to a people with a declared abhorrence of infinity. However, the Greeks' displeasure was surely tempered by the possibility that some of these numbers might indeed be given geometric descriptions. Upon further investigation, some irrational numbers came to be viewed as "better" than others, in the sense of being more intuitively appealing. The ancient Greeks subscribed to a philosophy of the necessity of the tangible nature of number and therefore admitted only those numbers capable of construction using the basic instruments of the geometer—straightedge and compass. Such *constructible* numbers include the square root of two, represented by the line drawn along the straightedge lying on a diagonal of a square whose sides have length one; they also include π, realized with the help of our trusty compass as the circumference of a circle with diameter equal to one.

Over time, this prejudice in favor of tangibility abated, and the explicit constructions embodied in geometry eventually gave way to implicit ones. A notion of number grew beyond that which is visually apparent to include that which we wish to be true. Soon, numbers were anticipated according to their ability to satisfy some mathematical property, a wish couched in an emotionless and universal language of equations, written as letters, numbers, and other much more exotic icons. The now familiar stalwarts of arithmetic—zero and the negative numbers—were at one time viewed as impossibilities because they lacked substance. But describe something, and it will come. For example, what of the number "minus one"? We envision a "quantity" which has the property that when added to two, it gives one, summarized by the symbolic sentence $x + 2 = 1$. Similarly, we dream of a number called "zero," defined by the property that when it is added to any number at all, it is without effect. No one would deny that these early *gedanken* numbers have turned out to be quite useful.

We can continue the implicit constructions that yield numbers like zero and the negatives, and arrive at a notion of number that includes instantiations of even more fanciful dreams: numbers defined in terms of relations among their powers. What of the number whose second power is equal to two? Or equivalently, what of a number whose square when added to minus two yields zero?* That is the now familiar square root of two. What of a number whose third power when added to twice its second power and then further diminished by seven gives zero?† Numbers like these are related to, but more general than, square roots, cube roots, and the like, and have the property that multiples of their powers can be balanced in such a way as to produce zero; they are called ***algebraic numbers.*** Numbers that do not fall under this umbrella are said to be ***transcendental,*** as the impossibility of their implicit description using only a finite number of powers somehow puts them beyond the apprehension of mortals. Although most real

*I.e., a square root of two is the number that when substituted for x in the expression $x^2 - 2$ produces 0. In other words, it is defined as a number that satisfies the equation $x^2 - 2 = 0$.

†I.e., a number that satisfies the equation $x^3 + 2x^2 - 7 = 0$.

numbers are transcendental (there are but a countable infinity of algebraic numbers within the uncountably infinite number of reals, leaving an uncountable number of transcendentals), it is notoriously difficult to establish the transcendence of any particular number. The proofs of the transcendence of e (by the Frenchman Charles Hermite in 1873) and π (by the German Carl Lindemann in 1882) are still hailed today as two of the great achievements of modern mathematics.

Within the classification of numbers as algebraic or transcendental, still more surprises await us. Algebra has opened gates onto vistas that seem beyond our intuition. Among the algebraic numbers occurs one whose definition seems quite natural: What of that number whose square is minus one? What kind of number could this be? Why should such a number be considered any different, any more "complex," than the analogously defined square root of two? In spite of their common algebraic origins, it is a square root of minus one that bears the name ***imaginary number.*** The real numbers and imaginary numbers together produce the collective of complex numbers.

In *An Imaginary Tale,* Paul Nahin tells the wonderful story of the intellectual birth of the complex numbers. In fact, having a number whose square is equal to minus one was the realization not of a vision but rather of the more prosaic desire to find a formula for those numbers that are described in terms of their ability to satisfy certain kinds of ***cubic polynomials.*** Such a number is said to be *algebraic of degree three* and has the property that there is some integer which, when added to some combination of multiples of its first, second, and third powers, gives zero. For example, you might look for a number such that three times its third power when added to minus two gives zero.* One such number is the cube root of two-thirds. Our goal is to arrive at a general scheme which, given the relation that we seek to satisfy, is able to give a formula for a number that satisfies the original cubic polynomial. This is analogous to the *quadratic formula* that some of us learn in order to solve relations involving at most the second power.

Nahin tells of the sixteenth-century mathematicians who posed problems and publicly challenged one another to "duels" for their solution. Through these public displays of intellectual acuity, winners

*I.e., a solution to the equation $3x^3 - 2 = 0$.

hoped to gain the patronage of the intellectually curious aristocracy, while losers suffered wounds only to the ego. One such mathematician-gambler-duelist, Girolamo Cardano (1501–1576), found himself forced to use, or at least manipulate, the square root of a negative number in order to solve a certain cubic polynomial. It was (in his words) "mental torture" to use such "sophistic" numbers.

These numbers that so tortured Cardano are instances of complex numbers. Euler was the first to symbolize the square root of minus one as ***i***. Other examples of complex numbers are $3 + 4i$, $-2 + 3i/2$, and πi. Most generally, a complex number is expressed as $a + ib$ where a and b are any real numbers. In this case a is the ***real part*** and b the ***imaginary part*** of the complex number $a + ib$. This augmentation of our mathematical language now enables the creation of a completely self-consistent arithmetic for this new and all-encompassing world of numbers. The totality of this "complex universe" of number, with its nesting of the real numbers within the complex numbers, and the rationals within the reals, and the integers within the rationals, and finally, the natural numbers within the integers, is pictured in Figure 9.

Despite their daunting name, complex numbers are actually the source of much simplification within mathematics. Problems that appear intractable in a restricted world which uses only real numbers magically decompose when cast into the larger complex setting. Rather than being complex, these are numbers of ease and simplicity, and in their utility for the sciences and engineering they are perhaps even more real than the real numbers. Even in their definition, they are more like life itself: partly what is and partly what we wish for, part real and part imaginary.

The Complex Plane

Real numbers live on the one-dimensional number line, but the independent nature of the real and imaginary parts of a complex number suggests a two-dimensional representation, like coordinates in the Cartesian plane. This setting is the ***complex plane,*** in which the labels *real* and *imaginary* replace the east-west and north-south, or x and y, of the Cartesian plane.

A portion of the complex plane is represented in Figure 10 with a

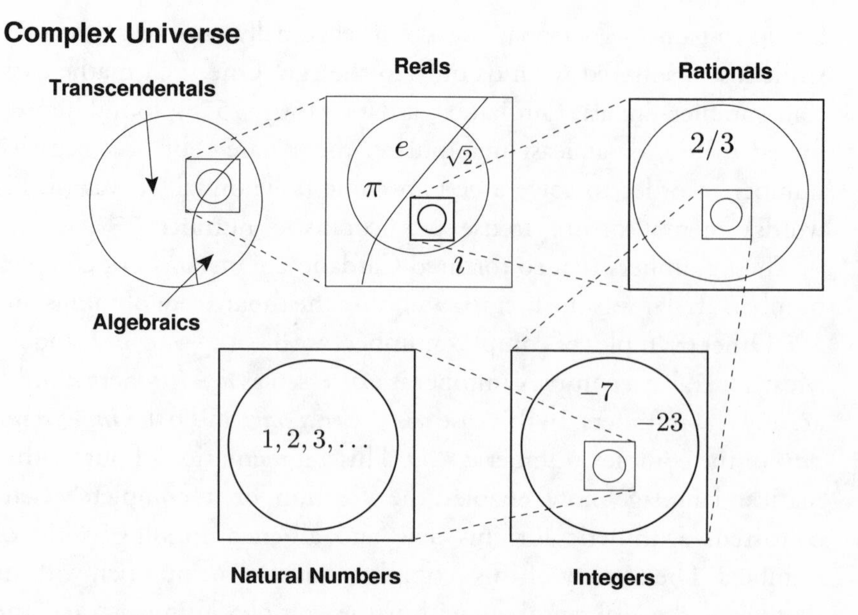

Figure 9. "Complex universe" of numbers. The complex numbers are either algebraic or transcendental, characterized, respectively, by being, or not being, a root of some polynomial. Examples of the former are i (a non-real number that is the root of the polynomial $x^2 + 1$) and 2, while e and π are examples of the latter. The complex numbers contain the real numbers (the next circle) of which some are irrational and others rational (those within the next circle). Finally, among the rationals are the integers and then among the integers are the good old natural numbers.

few representative numbers indicated. The ***imaginary axis*** runs north-south, while the ***real axis*** runs east-west. Thus, the real and imaginary parts of a complex number indicate the east-west displacement and the north-south displacement, respectively. The real axis is then simply our familiar old number line. *Purely real* numbers are just those complex numbers with imaginary part equal to zero and *purely imaginary* complex numbers are found on the imaginary axis. If both the imaginary and real parts are nonzero then the corresponding point is off the axes. For example, the complex number $1 + i$ is realized as the point one unit north and one unit east of the origin, while the complex number $-2 - 5i$ is two units to the west and five units to the south of the origin.

This complex picture was the invention of a little-known Norwegian surveyor, Caspar Wessel (1745–1818), who hit upon the idea while working on a topological survey of Denmark. He announced it

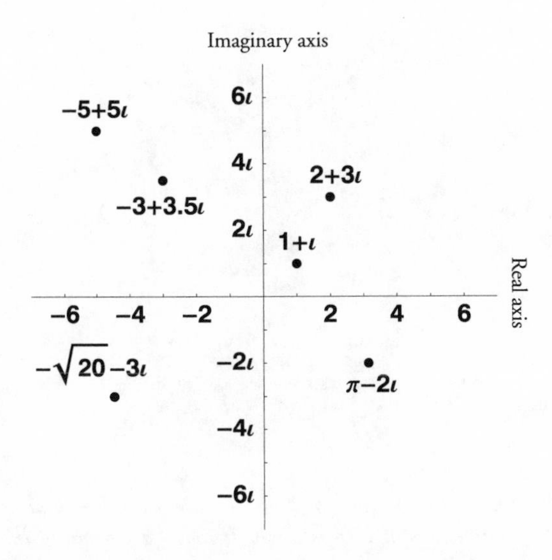

Figure 10. Portion of the complex plane with a few representative complex numbers.

in a paper presented to the Royal Danish Academy in 1797, and published it (in his only mathematical publication) in 1799.* This is a geometric representation of complex numbers that shows them to be mathematical quantities which have both magnitude (given by their distance from the origin) and 360 degrees of possible direction (not just the simple left-right of the number line), and which still obey all the usual laws of arithmetic. It is in this complex landscape that Riemann would develop a technique for studying the primes.

In his doctoral dissertation Riemann paved the way for extending the techniques of the calculus to the complex world. Newton's calculus of real numbers had made possible the dramatic progress of the physics of his day, describing the motion of macroscopic bodies, be they airborne apples or celestial systems. Riemann's complex calculus would turn out to be crucial for the next generation of physics; he

*Nevertheless, the concept of the complex plane is usually associated with the Swiss bookkeeper and amateur mathematician Jean Argand (1768–1822). His short, self-published, and anonymous pamphlet of 1806, *Essay on a Manner of Representation of Imaginary Quantities through Geometric Constructions,* gave a thorough presentation of this Cartesian or "rectangular" representation of complex numbers. This pamphlet had a fairly wide circulation, and it was only after another mathematician republished the results in a professional journal, and asked the original author to step forward, that Argand became known for this contribution.

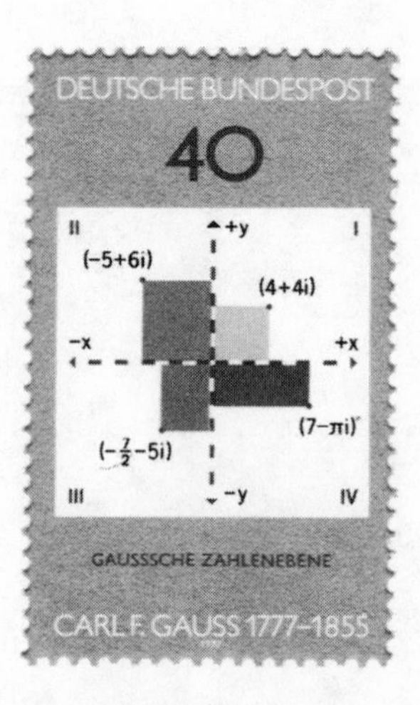

Figure 11. Stamp (of the former West Germany) commemorating the two hundredth anniversary of Gauss's birth. It shows a portion of the complex plane with a few illustrative points. (Image from J. Kuzmanovich.)

anticipated the future needs of science by constructing the mathematics fundamental to understanding, modeling, and then manipulating electromagnetic phenomena. Riemann's contributions to the invention of complex analysis continue to be among the most important mathematical tools necessary for the design of technologies including radar, microwave ovens, CD players, and digital cameras. But the complex calculus did more than provide new possibilities for home appliances; it also transformed the study of prime numbers. By extending the work of Euler and Dirichlet to consider a series of powers in which the exponent is a complex number, Riemann would give birth to the ***Riemann zeta function*** and thereby create a tool that could enable mathematicians to settle once and for all the question of the growth of the count of the primes.

Complexifying Euler: the creation of the Riemann zeta function, a number theorist's PDA

The harmonic series served as a template for the investigation of properties of the primes by Euler (who considered only the sum of reciprocals of integer powers) and Dirichlet (who used the whole real line as potential exponents). We can think of this as using the harmonic series as the guts of a mathematical machine designed to study the primes: a *prime distribution analyzer,* or PDA, for the curious number theorist. This PDA does not organize your calendar (unlike its office assistant cousin, the other PDA, the personal digital assistant); rather, the mathematical PDA will help bring organization to the investigation of primes. It will be the means by which we describe Riemann's invention of the zeta function—his tool for describing the accumulation of the primes.

The original Eulerian PDA had but a single knob. Its discrete natural number settings indicate the exponent currently under investigation. With the knob set at one, we get the harmonic series—that is, the sum of the integer reciprocals raised to the power one—and an attached screen would register the toppled figure eight that is the sign for infinity, indicating the unbounded growth of the harmonic series. With each subsequent clockwise click of the dial, we get an increment in the exponent of choice. Each setting greater than one produces a number on the screen, indicating the finite bound or limit of each of these other infinite series. A single clockwise click from one to two produces the results of summing the reciprocals of squares. Two clicks give an exponent of three and the consideration of sums of the reciprocals of the cubes. Each of these finite outcomes (as compared with the infinite sum of reciprocals of the primes) reflects what we have already learned: that the primes are more densely distributed than any collection of powers.

Dirichlet's version improves upon the original PDA design by permitting clockwise turns of the dial of any amount, enabling the evaluation of the series for exponents between the integers. This capability of fine-tuning in Dirichlet's PDA revealed new possibilities for understanding the primes; in particular, it allowed Dirichlet to prove that almost any arithmetic progression contains an infinity of primes.

Riemann updated the PDA to permit complex exponents: "The Riemannian PDA—a new and improved PDA for today's complex world!" This will be Riemann's zeta function.

The Riemannian PDA: Riemann's Zeta Function

Whereas Dirichlet had located the harmonic series as a single series in the one-dimensional number line–like world of real exponents, Riemann's facility and familiarity with the calculus of complex numbers must have suggested to him that much could be gained by seeing the harmonic series as only one instance in the wider and freer complex world, a world with two dimensions of choice as described by its independent real and imaginary parts. Riemann considered the possibility of infinite series of integer reciprocals raised to powers that were more than integers, more than real numbers. He permitted exponents that were complex numbers. He thus replaced Dirichlet's single-knob PDA with a two-dial version reflecting the independent settings of real and imaginary. His hope, and indeed what would turn out to be true, was that the aggregate of these outputs would enable mathematicians to go boldly where they had not yet been able to go in their explorations of the distribution of primes.

Let's take a moment to look under the cover of the new Riemannian PDA. It requires the calculation of numbers raised to powers that are complex numbers. What is this?

Complex Power

Complex exponentiation is often considered a mysterious operation—what could it mean to raise a number to the square root of minus one-th power? The definition is a classic example of what could be the mathematician's credo: Not only is it better to teach people to fish than give them fish, but it is best to teach people to fish using the most basic of tools, say an arm for a fishing pole, a strand of hair for a line, and a fingernail for a hook. First principles above all.

Recall that one way to compute the number e is as the amount of money returned after one year with a $1 initial deposit in a bank that gives 100 percent annual interest, compounded continuously. Using this description it is possible to define the value of e raised to any complex power. In this case it will be the amount of "money" returned on a $1 deposit by the same continuously compounding bank, but with an annual rate of return that is a complex number.

For example, suppose that in the complex plane there is a complex bank from which we expect to receive complex money. In particular, suppose that a new savings plan allows for an annual return of i dollars on an investment of $1, compounded once per year. Thus at the end of the year we would receive i dollars in interest, and so a total return of $1 + i$ dollars at the end of the year. If the interest is compounded twice, we will receive (in principal and interest) $1 + i/2$ dollars at midyear, which would then generate another $i/2$ times that quantity, or $i/2 - 1/4$ dollars in the next half of the year (this is due to the fact that i squared is minus one) for a year-end total of $3/4 + i$ dollars.*

Continuing in an analogous fashion,† if the interest is compounded continuously, we will receive an amount of money that is equal to e raised to the power i. Through the ability to express any other number as a power of e (in particular, according to its definition, any real number is equal to e raised to its natural logarithm), it is then possible to calculate the result of raising any number to a complex power. For example, suppose you wanted to know the value of three raised to the power i. Since three is e raised to a power that is a little more than one, this would be the same as raising e to the power i, and then raising this number to that power that is a little more than one.‡

When this process is followed using a return of πi dollars, we obtain one of the most surprising and beautiful of mathematical facts: that e raised to the power πi is equal to minus one. The equation

$$e^{\pi i} + 1 = 0$$

is a famous mathematical haiku, tying together several fundamental mathematical symbols: 0, 1, e, i, π. This equation is called *Euler's identity*, and Euler considered it so magical and beautiful that he used it for his epitaph.

So Riemann's PDA takes as input a complex number, and computes

*Compounding twice means that at the end of the year we will receive $(1 + i/2)^2 = 1 + i + (i/2)^2 = 1 + i - 1/4 = 3/4 + i$ dollars.

†Compounding N times per year means that at the end of the year we would receive $(1 + i/N)^N$ dollars. Compounding continuously means letting N get bigger and bigger and seeing if the numbers that we get for these large values of N approach a limit, in the same way that Zeno's walk across a room gets as close as we like to the other side.

‡In other words, since 3 is equal to $e^{1.0986\ldots}$, then $3^i = (e^{1.0986\ldots})^i = e^{i(1.0986\ldots)}$, which we calculate using the methods just explained.

the putative sum of the reciprocals of all the natural numbers, each raised to this common complex exponent. In the mechanistic view of mathematical function as a machine that takes in numbers, transforms them according to some rules, and spits out new numbers, Riemann's PDA is a machine of machines, comprising some that exponentiate and others that add. Rather than call this the function-that-turns-complex-numbers-into-prime-distribution-data (as he might have done, given the freewheeling word formation of the German language), he called it the *zeta function,* simply because he used this letter of the Greek alphabet to notate it. Accordingly, we call it ***Riemann's zeta function.***

Dirichlet's previous work on arithmetic progressions used only real exponents greater than one. By using the harmonic series template to define the zeta function Riemann was able to make sense of exponents that were complex numbers with a real part bigger than one (remember that complex numbers have a real part and an imaginary part). But Riemann needed more. To use his knowledge of complex analysis fully, he needed a way to make the dial of his PDA go all the way. And to do this, he realized that he knew another, shorter way to say "take the sum of the reciprocals of all natural numbers raised to a complex exponent." That is, he knew a different way to get those outputs on the Riemannian PDA. A pocket calculator and home computer give the same results for adding two numbers, but the home computer is much hardier, with a much wider range of applicability. Similarly, Riemann found one last update on the first version of his PDA that worked for all the complex numbers with a real part greater than one and also worked for all the other complex numbers. This version 2.0 of the Riemannian PDA is the *integral form of Riemann's zeta function,* and it defines the Riemann zeta function for the entire complex plane.*

To use the Riemann zeta function to study primes, one more step is

*In doing this, Riemann found a function, defined on the entire complex plane, that took the same values as the modified harmonic series when used with complex inputs with a real part greater than one. This is called an *analytic continuation* of the series, a name that comes from the notion of having continued the definition of the function past its original domain to a function that now can be used everywhere in the complex plane.

needed. Both Euler and Dirichlet found that the true utility of the harmonic series for these investigations was realized when they considered not just the simple accumulation of the sums of the reciprocals, but rather the logarithm of these sums. So Riemann did the same with his complex version, effectively taking his complex PDA and plugging it into a new machine which would transform the Riemann PDA output and spit out the logarithm (perhaps a basic logarithmic transformer, or BLT).

When Riemann did this—i.e., considered the logarithm of the Riemann zeta function—a miracle occurred: had this last component in the sequence of machines been hooked up to a set of speakers, the result would have been music to his ears. For with this last transformation, the primes began to "sing." In the logarithm of the zeta function, Riemann saw glimmers of something that looked like music or, rather, the mathematical representation of music that is a Fourier transform, one of the key techniques of ***Fourier analysis,*** the last ingredient in our understanding of the Riemann hypothesis.

The Music of the Prime Powers

Fourier analysis is the mathematics of wave phenomena, providing the tools to analyze and understand, for example, the gentle ripples caused by a stone tossed into the middle of a quiet pond, the disturbances in the air caused by the rat-tat-tat of a drumhead, or the placing of a call on a cell phone.

The subject acquires its name from the mathematician Jean-Baptiste-Joseph Fourier (1768–1830), who took on the problem of developing a mathematical analysis of waves in order to derive a theory of heat transfer. He published his discoveries in 1807 in a memoir, *On the Propagation of Heat in Solid Bodies.*

Heat, in all its various forms, physical or emotional, seems to have been a general theme in Fourier's tempestuous life. In the politics of France during his day, he was continually jumping between the fire and the frying pan. Fourier narrowly escaped the guillotine during the Reign of Terror following the French Revolution; and as Napoleon's power and favor waxed and waned, he rode with its flow and retreated with its ebb.

When Napoleon left France to invade Egypt, Fourier signed on to accompany him as a scientific adviser. Fourier was an able administrator as well as a first-rate scientist. While in Egypt Fourier founded the University of Cairo and led the archaeological studies commissioned by Napoleon. The scientific achievements of the expedition were collected in *Descriptions of Egypt* (1811), to which Fourier contributed a highly regarded general history of Egypt. Upon Napoleon's return to the Continent, Fourier was appointed prefect of the Isère region in the French Alps. His chief administrative accomplishment during his tenure was the construction of the first road over the Alps between Turin and Grenoble.

Fourier's stint in Egypt seems to have left a tremendous impression. According to at least one historical note, he spent the latter years of his life so convinced of the healthful powers of heat that he swaddled himself in cloth and lived and worked in overheated rooms. This leaves us with the image of a mummylike Fourier, working away feverishly in a hot and dry study, driven to invent the mathematical tools capable of revealing the secrets of heat flow, and in turn, the key to the astonishingly well-preserved sarcophagi that he saw in Egypt.

Between 1804 and 1807 Fourier developed a theory of heat flow based on Newton's law of cooling, which states that the movement of heat between two bodies is proportional to the difference in their temperatures. To take this simple law and transform it into a mathematical law capable of representing the point-to-point transfer of heat which occurs as a piece of metal is warmed by a small flame is a considerable intellectual leap. Yet this is what Fourier accomplished, writing down the fundamental ***partial differential equation*** that today is called the *heat equation.*

The heat equation, like the equations that define complex numbers, gives an implicit description of the flow of heat, and so is a first step in understanding the phenomenon. Rather than attempt to describe how much heat is in any given location, it seeks instead to describe how the heat changes over time. This explicit description of the associated dynamics, a description that might be pictured as a changing temperature map of a city or region, is the solution to this partial differential equation.

A specific problem addressed in Fourier's memoir is the heating and cooling of the earth. Local temperature fluctuations can be understood as cycles within cycles, a terse numerical summary of the interweaving of the diurnal and yearly cycles that result as the earth circles the sun, from apogee to perigee and back again, while simultaneously rotating on its axis, turning any part of its surface toward and away from the sun so that day turns into night and night turns again into day. In order to understand the concomitant cyclic variation in temperature, we need to have at hand the analogous fundamental mathematical cycles or, as they are also called, ***periodic functions.*** These basic periodic functions are ***sinusoids,*** the sources of the sinuous graphs traced out by the sines and cosines of high school trigonometry. Charted on graph paper they give ideal waves forever moving along an east-west axis that divides the infinity of equispaced crests and troughs of a never-ending mathematical sea serpent. These crests can be close together or far apart, corresponding to waves of greater or lesser ***frequency.*** The crests can be very tall or quite small, tsunamis or ripples on a pond, corresponding to waves of greater or lesser ***amplitude.***

Recast in the more familiar sonic setting, each of these waves corresponds to one of an infinity of possible tuning forks, each with tines that vibrate at a characteristic frequency (pitch), but of varying displacement (amplitude) which depends upon the force with which it is struck. The simultaneous or nearly simultaneous striking of several tuning forks results in a ***superposition*** of the sonic waves initiated by each of the tuning forks. The decomposition of any sound into its basic "tuning fork" sounds or frequency components is the Fourier analysis of the sound, and the mathematical operation that effects it is called the ***Fourier transform.***

Using such basic sinusoidal elements, Fourier wrote down the first *Fourier series,* which in this case is a sum or superposition of basic waves of temperature that combine to create the cyclic ebb and flow of heat on the planet. The lack of detail in his memoir of 1807 caused it to languish in the hands of its reviewers, a committee consisting of two of the most famous mathematicians of the day (Joseph-Louis Lagrange and Pierre-Simon Laplace) and two of the most famous physicists (Gaspard Monge and Sylvestre Lacroix). Four years after the submission of Fourier's work the committee had still not accepted it and, moreover, even doubted its veracity. As a spur to the scientific community (undoubtedly motivated by the hope of finding a more worthy manuscript than Fourier's) a competition for papers on the mathematics of heat conduction was announced. But after deliberation (and despite some remaining dissent) the prize was finally given to Fourier for an updated version of his paper—although it was not until 1822 that his prize essay was eventually published. This was only one of the many honors accorded to him over the last years of his life, during which the furor surrounding his work and academic appointments cooled. Ultimately, Fourier received his deserved recognition (elections to the Academy of Sciences as well as the Académie Française); and he lived out his later years in comfort, respected for a lifetime of scientific and social achievements.

Today, Fourier series are ubiquitous in the applied sciences. Their importance in the analysis and exploitation of electromagnetic waves indicates the central role of Fourier analysis in all things digital, perhaps most notably digital audio engineering and digital image processing, the former the science of compact discs and MP3s, the latter the mathematical foundations enabling digital cameras, MRIs, and CAT scans.

Dirichlet had been among the first to look closely at Fourier's work, and thus Riemann's study of Fourier analysis is another example of Riemann's following in Dirichlet's wake. Riemann began his investigations after completing his doctorate in complex analysis. This was the time in Riemann's career when he was required to submit additional research in order to rise to the rank of *Privatdozent,* a title and certification that would allow him to be an unpaid lecturer at Göttingen, with the right to take on students who would pay a small amount of money in return for the privilege of attending lectures. This was a terribly onerous way to make a less than mediocre living. Even at Göttingen, a hotbed of mathematics, it was unlikely that Riemann would have had enough students to provide a living wage.

Like Dirichlet, Riemann investigated the question of what sorts of phenomena could be described by the steady beat of Fourier analysis, and it was his familiarity with such issues that proved critical for the future of number theory. For Riemann was able to recognize that implicit in his zeta function was (in his words) the "thickening and thinning of primes" that marked the irregular beat of prime and composite.

Riemann's explicit formula and the zeros of the zeta function

Through Fourier-analytic eyeglasses Riemann was able to see encoded in the zeta function an explicit formula for counting the number of primes less than a given magnitude. He saw a way to express the counting of the primes as a term-by-term accumulation of fine detail slowly etched upon coarse behavior. An initial approximation comes from a long and leisurely wave, with more and more detail added by waves of higher frequency.

The basic waves that describe music are identified by their *frequency.* This single real number ranges from low to high as sound moves from bass to treble. In effect, to describe any sound it is enough to give a recipe that lists each possible frequency, and how much of it should be included—this is the *amplitude,* and thus reflects the volume at which the sound should be played. Alternatively, each of the pure sonic waves is a variation on that perfect sinuous sine or cosine wave that the hot and bothered Fourier analyzed so deeply. Designating a particular frequency means detailing how much to either squeeze together or spread apart the successive peaks, and the amplitude determines how high the peaks will fly or, equivalently, how low the troughs will go. Thus any sound is described as a *superposition* or sum of a potentially infinite series of waves, each described by two numbers: one for frequency and one for amplitude.

Similarly, the accumulation of primes will be a sum of fundamental "tones," but tones that are variations of slightly more complicated waves. In this way the function that keeps track of the accumulation of the primes would be realized like a symphonic masterpiece: built from bass to treble, long low tones of the kettledrums and bass drums rolling like thunder below the deep overtones of the bass, cello, and oboe, over which ride the increasingly ethereal clarinets, violas, and violins, and culminating with the tremulous trills of the flutes and piccolos. Ultimately, at some infinitely distant horizon, their simultaneous sounding gives a perfect replication of the prime distribution function.

Sound waves are described using two real numbers indicating their frequency and amplitude. Similarly the primal tones are described using two real numbers, which are further combined into a single complex number using these real numbers as its real and complex part.

The elemental primal resonances are described by very special complex numbers, and this is where Riemann's zeta function makes its magical appearance. For the complex numbers that delineate the fundamental tones whose symphony is the accumulation of primes are precisely those complex numbers that bring Riemann's zeta function to zero. They are the tunings on the dial of the Riemannian PDA that cause its readout to flatline. These settings are called the *zeros of Riemann's zeta function,* or simply the ***zeta zeros.***

Riemann determined that there are an infinity of such settings of the dials, an infinity of complex numbers each of which simultaneously marks a place where Riemann's zeta function vanishes while cuing a next number-theoretic tuning fork to join the growing chorus of waves sounding out the accumulation of the primes. These are "primal waves," whose bowing and bending contribute to that accumulation. Two of these primal waves are shown in Figure 12. Notice that in these waves, unlike basic sinusoidal waves, the crests grow and are farther apart as the wave goes farther out—a fact that will be very useful in describing the growth of the primes, which generally have longer and longer gaps between them as they get larger.

The superposition of these prime fundamentals gives the real behavior of the primes as they surface in the integer sea. Each zeta zero is less a void than a renewal, signifying the entrance of some new detail into the growing description of the song of the primes.

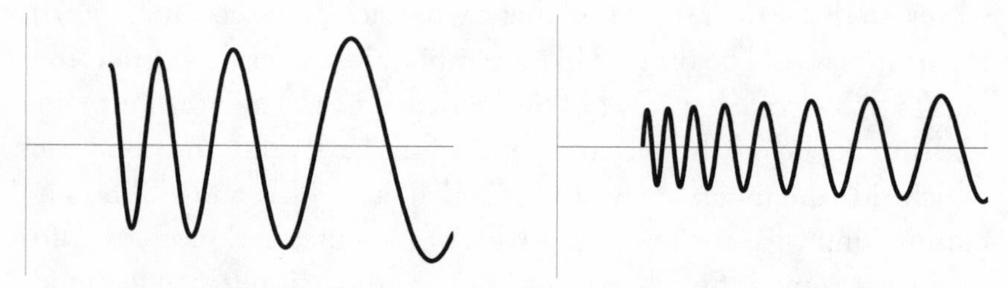

Figure 12. The first (left) and fourth (right) basic primal waves. The former represents the low bass tone of the primes, or the first primal partial; the latter is the fourth primal partial.

The (improving) estimates of the count of the primes between 190 and 230 are shown in Figures 13, 14, and 15, achieved by using, respectively, the first five, fifty, and, finally, five hundred primal waves, whose characteristics are encoded in respectively the first five, fifty, and five hundred zeta zeros. Using the first five zeta zeros brings out only the coarsest characteristics of the jagged graph (Figure 13). If the first fifty zeta zeros are used, more of the shape comes out (Figure 14), and finally with the first five hundred zeta zeros the steps in the graph of the true count are reproduced almost perfectly (Figure 15).

Why the Zeros?

At first glance it seems like a mystery. Why would the zeta zeros carry the information for understanding the count of the primes? We must

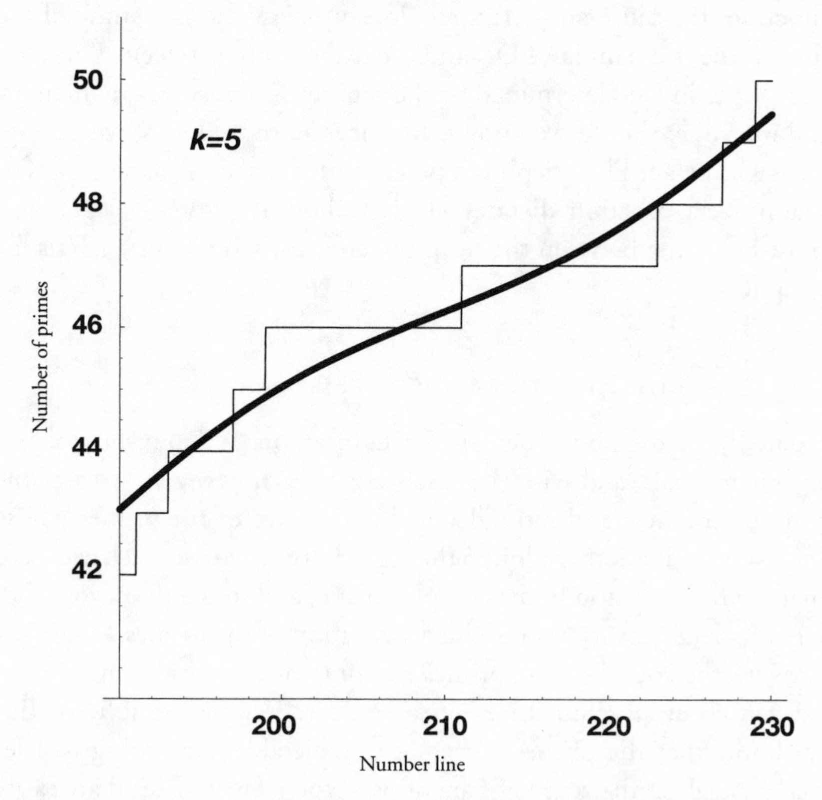

Figure 13. The curved line shows the approximation to the prime distribution function counting the number of primes between 190 and 230 as a superposition using only the "primal waves" determined by the first five zeros of Riemann's zeta function.

remember that although the fact that the harmonic series diverges to infinity is one way to prove the infinitude of the primes, the manner in which the harmonic series becomes infinite is what is of concern in order to understand the rate of this unbounded accumulation. This is the reason that Riemann has taken us to the complex plane: to understand the nature of the infinity of the primes through the study of his zeta function. These days, theoretical physicists believe that our mundane four dimensions of time and space require an understanding of a grander ten-dimensional physical world. Analogously, Riemann has told us that our understanding of the way in which the primes are strewn throughout the one-dimensional number line will require a new understanding of phenomena in a world of two dimensions (as represented by the real and imaginary parts): the complex plane.

The phenomenon in question is the behavior of Riemann's zeta function, the cause-and-effect relationship between the tuning of the dial on the Riemannian PDA and the readout on its screen. This relationship is in fact determined by the zeta zeros. For such a mathematical machine, should we know where it is zero, we know everything. These zeros act like telephone poles, and the special nature of Riemann's zeta function dictates precisely how the wire—its graph—must be strung between them. In the end, truly is all our analysis for naught.

THE RIEMANN HYPOTHESIS

Excitedly we tune into Riemann's zeta function. We spin the real and imaginary dials, and with the discovery of each new zero we are able to fashion a new and more detailed basic wave of the primes whose inclusion adds another kink to the jagged primal curve. Each zeta zero marks a new location in the complex plane, and so their positions dictate the detailed variation in the distribution of the primes.

Were the sound of the primes a simple one, a steady rather than uneven accumulation, these zeros would reflect that simplicity. But we know that the primes are not so predictable. According to Riemann's analysis the zeta zeros are of two types. The first kind are easily derived from known properties of Riemann's zeta function. They are the negative even integers, neatly arranged on the west side of the real

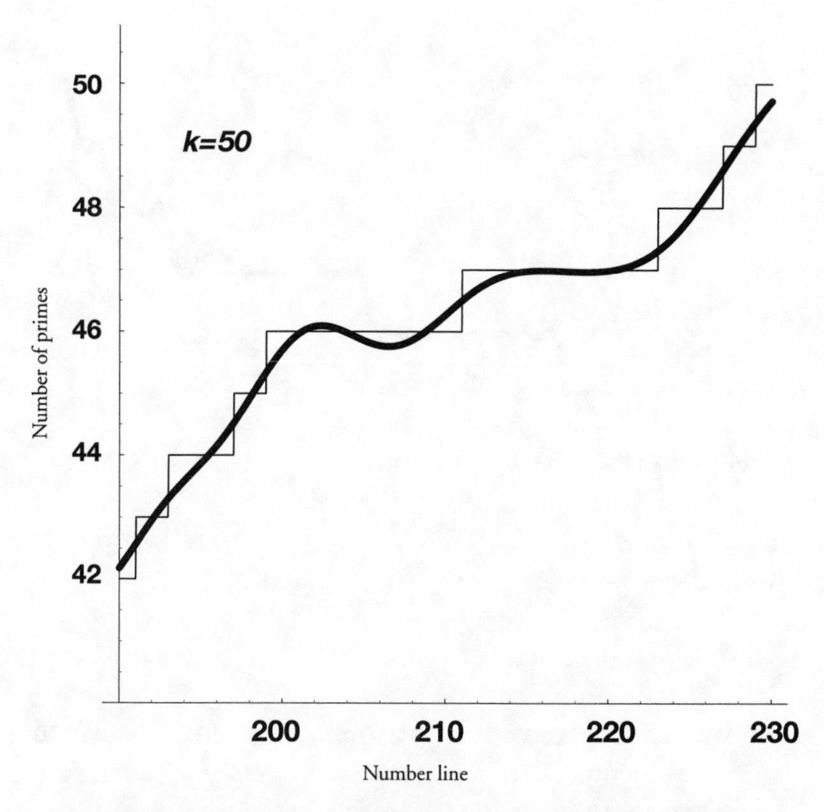

Figure 14. Here the thick curved line shows the approximation to the prime distribution function counting the number of primes between 190 and 230 using the primal waves determined by the first fifty zeros of Riemann's zeta function.

axis of the complex plane. The ease with which they were determined has led to their being called the ***trivial zeta zeros.*** A great deal is known about these zeros.

It is the other zeta zeros, the ***nontrivial zeta zeros,*** which are the source of all the mystery. Without too much difficulty Riemann was able to constrain their positions to within one infinite vertical highway running through the complex plane. This is the famous ***critical strip,*** a region of infinite extent within the complex plane. The critical strip is bounded on one side by the north-south (imaginary) axis and on the other by a line parallel to the imaginary axis but one unit to the right, so that it is also perpendicular to the real axis but goes through the point corresponding to the number one.

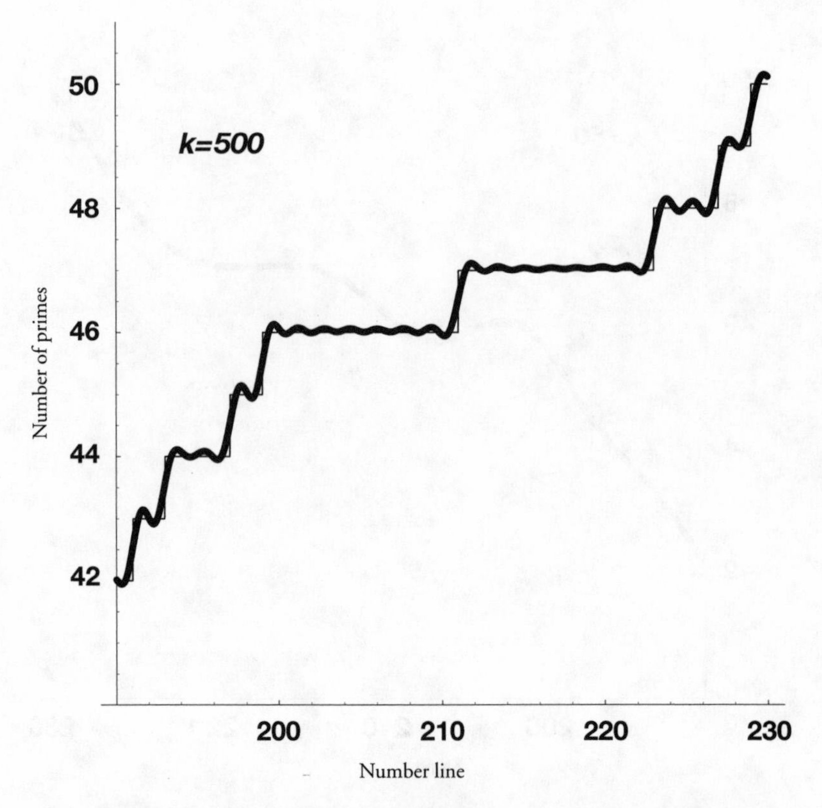

Figure 15. The thick line here is the approximation of the prime distribution function counting the number of primes between 190 and 230 using the primal waves determined by the first 500 zeros of Riemann's zeta function. It is almost a perfect fit.

Zeta zeros in the critical strip can be described according to how high up in the strip they are found. This is called either the ***height*** or ***level*** of the zeta zero. On the one hand, Riemann estimated that the number of zeta zeros in the critical strip below a given height was approximately equal to the height times its logarithm. On the other hand, he estimated that the density of zeros near a given height was about equal to the logarithm of the height.* Thus, if we count off the zeros one by one according to height, he estimated that the height of a zero would be approximately equal to its rank (i.e., at what place in

*More specifically, Riemann conjectured that the number of zeta zeros at height at most T was about $(T/2\pi) \times \log(T/2\pi)$, while the density of zeta zeros was approximately $\log(T/2\pi)$.

the listing the zeta zero occurred) divided by the logarithm of its rank. In this way the zeta zeros seem to be like a strange mirror of the primes themselves. Their behavior seems to reverse the form of the primes as conjectured by the Prime Number Theorem, which as we have seen asserts that the number of primes less than a given amount is that amount divided by its logarithm, and that the size of a particular prime is approximately its rank in the order of the primes times the logarithm of that rank. Thus while the primes occur farther and farther apart as we move farther and farther along the number line, the zeta zeros become more and more closely packed together the farther along we advance up the critical strip. And the density of the one is approximately the reciprocal of the density of the other.

The goal of Riemann's paper was the derivation of his explicit formula, and with his eyes on this prize, and perhaps in his haste to return to his other research, Riemann did not give a much more detailed description of the nontrivial zeros. Regarding these nontrivial zeros Riemann knew much more, and all this forms the basis of his famous hypothesis. In particular, he believed it "very likely" that the zeros were not simply within the critical strip, but that in fact they split the critical strip, all falling on the ***critical line,*** the north-south line defined by those complex numbers whose imaginary part is equal to one-half. This is the ***Riemann hypothesis,*** and the quest for its resolution lives on today as perhaps the most important unsolved problem in mathematics. In typical Riemannian understatement he wrote:

> *I have put aside the search for such a proof after some fleeting attempts because it is not necessary for the immediate objective of my investigation.*

We know today that these "fleeting attempts" are actually pages of calculations.

Encoded in the Riemann hypothesis is the precise detail of the accumulation of the prime numbers, and its solution is today linked not only to an understanding of the natural numbers, but also, as we will see, to nature itself. At the time of its statement, however, it seemed to the man who posed it a triviality, which, were he to have enough time, he could and would surely settle.

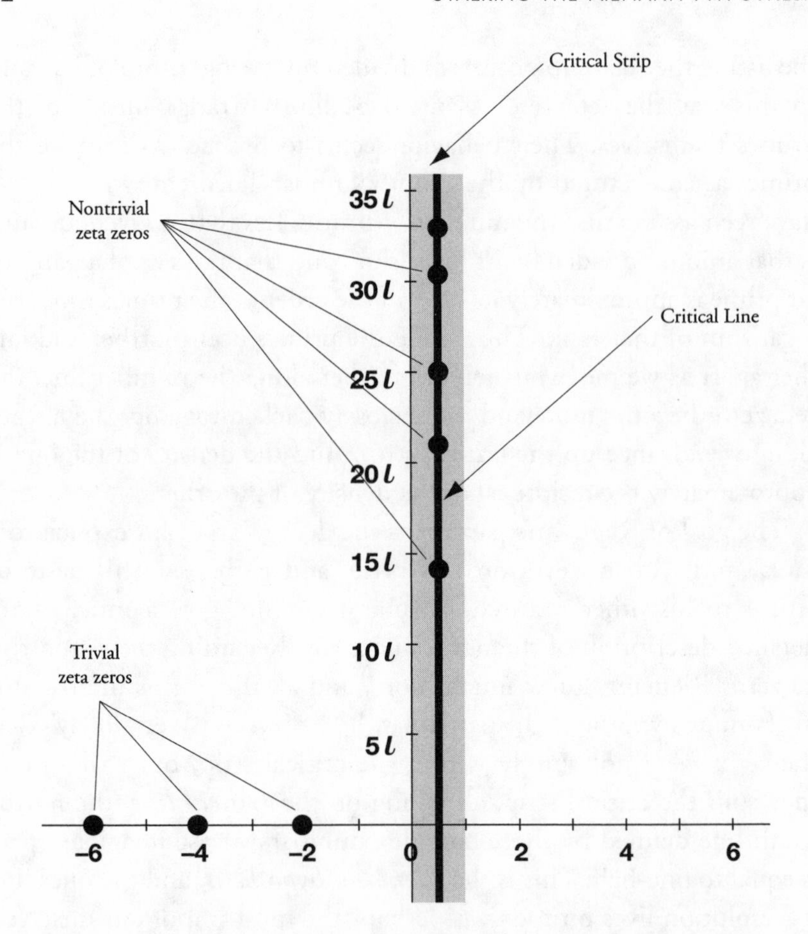

Figure 16. The complex plane in the region nearby the origin. The first few trivial zeta zeros (−2, −4, −6) are indicated on the horizontal (real) axis. The critical strip contained herein is indicated by the shaded region. The critical line splits the critical strip and consists of all complex numbers with real part equal to one-half. The Riemann hypothesis conjectures that any nontrivial zeta zero lies on the critical line. This statement has dramatic implications for the determination of the precise distribution of the prime numbers. The large dots on the critical line indicate the positions of the first few nontrivial zeta zeros which were computed by Riemann.

A QUIET ENDING

Unfortunately, time was a luxury that Riemann did not have. After delivering his mathematical homily in Berlin, he walked away from number theory and returned to the study of mathematical physics. In

1862 he developed pleurisy, which soon led to tuberculosis. Riemann's health, never strong, grew steadily worse, until in the summer of 1866, while on a holiday near Lago Selesca in Italy, he died. His biographer and friend, the mathematician Richard Dedekind (1831–1916)—who would extend Riemann's work even further by creating the Dedekind zeta function—records those last moments:

> *The day before his death he worked under a fig tree, his soul filled with joy at the glorious landscape around him. . . . His life ebbed gently away, without strife or death agony. . . . The gentle mind which had been implanted in him in his father's house remained with him all his life, and he served his God faithfully, as his father had, but in a different way.*

With Riemann's death, mathematics and science were robbed of one of their great original thinkers. His legacy was a new mathematics upon which a new physics of relativity would be built, as well as a deeper understanding of the primes, embodied in the mystery that is the Riemann hypothesis.

Like Moses, Riemann delivered from on high a terse collection of symbolic sentences that would guide his descendants for years to come; and again like Moses, he too would be unable to accompany his followers on their new quests. Mathematicians of the next generation would uncover the import of the Riemann hypothesis. They would find embedded in this "very likely" assertion a measure of the exactitude of the (still unproved) Prime Number Theorem, a quantification of the degree to which the exact count of the accumulation of primes differs from the asymptotic estimate. Years later, mathematicians, physicists, and computer scientists would connect the resolution of this conjecture to phenomena spanning the scientific universe.

But this was all to come. When first announced, the Riemann hypothesis received a measure of attention and interest commensurate with the quiet way in which it had been proposed. This problem of determining the exact location of the zeros of Riemann's zeta function seemed of secondary importance, viewed as a technical nicety, disposed of easily once the details of the long anticipated Prime Number Theorem were nailed down. Since Legendre and Gauss first broached its possibility, the achievement of the Prime Number Theorem had been the Holy Grail of number theory. With the help of Riemann's explicit formula it now seemed within reach. Surely, this goal overshadowed any need to confirm Riemann's rather technical and "very likely" hypothesis.

As we will see, a proof of the Prime Number Theorem was indeed just around the corner, accompanied by the first tentative but promising steps toward the resolution of the Riemann hypothesis. Ironically, this real, rigorous progress would come as something of an anticlimax, appearing in the wake of an announcement of a solution to the Riemann hypothesis. It was, however, an announcement whose justification, to this day, remains unknown.

7 *A Dutch Red Herring*

In 1885 the Dutch mathematician Thomas Stieltjes (1856–1894) announced that he had discovered a proof of the Riemann hypothesis. In a short article in the journal of the French Academy of Sciences, *Comptes Rendus,* Stieltjes wrote that, as regards Riemann's assertion concerning the location of the zeros of the zeta function, "I have managed to put this proposition beyond doubt by means of a rigorous proof."

Stieltjes's note contained neither a proof nor any serious attempt at one. At the time of his announcement the Riemann hypothesis had been unsolved for just over twenty-five years. It was a mathematical youngster in comparison with such open problems as the Prime Number Theorem, which was then almost 100 years old (and still unproved), or Fermat's Last Theorem, which was then nearly 200 years old. Thus the intrinsic difficulty of Riemann's remark had yet to be determined.

An announcement of a result, unaccompanied by a proof, was not so unusual, and the mathematical community awaited the publication of the implied sequel. There was no a priori reason to doubt the claim of the young Stieltjes. He was a well-known mathematician, albeit with a nonstandard academic pedigree. As a university student, he had been so keenly interested in mathematics that he neglected all other studies in order to devote himself single-mindedly to a self-guided mathematics curriculum whose reading list contained only the original works of masters like Gauss. As a consequence, Stieltjes repeatedly failed his exams at the Polytechnical School of Delft and was forced to leave without a degree. His father, a well-known civil engineer and politician, was able to secure for him a position as "assistant for astronomical calculations" at the Leiden University observatory.

The observatory's director allowed Stieltjes the freedom to continue his mathematical investigations, but in spite of a growing body of significant accomplishments in analysis and geometry, Stieltjes's lack of a degree proved to be a serious stumbling block to obtaining a regular university position. Ultimately, some backroom politicking by Charles Hermite (the same Hermite who proved the transcendence of the number *e*) eventually obtained for Stieltjes an honorary degree from Leiden, and in 1889 he was finally appointed to a professorship at the University of Toulouse, a position that he held for the rest of his days. Stieltjes is remembered as one of the great mathematical originals, and his legacy is honored by the University of Leiden's Thomas Stieltjes Institute for Mathematics, a center designed to promote interactions between pure and applied mathematics.

An inverted approach to the Riemann hypothesis

Stieltjes's approach to the Riemann hypothesis was in keeping with his idiosyncratic nature. Like Euler, Stieltjes turned the problem on its head and found a reformulation of the Riemann hypothesis in terms of an analysis of the ***reciprocal*** of Riemann's zeta function.

The reciprocal of a number is a familiar concept: it is the effect of computing "one over" the original number. The extension of this "one-overing" operation to a mathematical function is straightforward, adding one more module to the numerical production line. It thereby creates a new machine from the old by dividing one by the output from the original. This slightly enhanced mathematical function is the reciprocal of the original. For example, the reciprocal of the squaring function is simply the mathematical function that takes a number, squares it, and then takes this square's reciprocal. So it turns two into one-quarter, one-third into nine, and negative one into one.

On the surface, the reciprocal of Riemann's zeta function might not seem any easier to analyze than the zeta function itself. In tacking on one little operational step, it would hardly seem that we have created a situation so different from the original that the Riemann hypothesis would suddenly become obvious, or even approachable. But Stieltjes's genius lay not simply in the consideration of this different mathematical function but also in his ability to find a way of expressing it that revealed a new path to the Riemann hypothesis.

Stieltjes's description of the reciprocal of the zeta function, like a beautiful passage of poetry, shed a new and surprising light on what was becoming an increasingly familiar subject. He used a formula developed by Euler for the reciprocal of the zeta function. Like the zeta function itself, Stieljes's formula also uses the harmonic series as its starting point. But rather than adding all the reciprocal powers of the natural numbers, it requires adding some terms, subtracting others, and completely neglecting still others. The rule according to which the power of a given number is either added, subtracted, or ignored is known as the ***Möbius inversion formula,*** after the Austrian mathematician A. F. Möbius (1790–1868).

Although in our story Möbius plays the role of a number theorist, he is probably best known today for his contributions to topology, in particular for his invention of the famous *Möbius band* or *Möbius strip*. This geometric object is most familiarly made by taking a paper loop, cutting it once, inserting a half twist, and then reconnecting the ends. With this minor modification we achieve a mathematical party decoration that is a single surface (as opposed to the inside and outside or top and bottom of the original paper loop): an ant walking its "equator" will be able to traverse in one single lap what had previously been two opposite sides of a loop. This single surface is significant in that it is *nonorientable;* that is, our meandering ant will not be able to tell up from down, since a half lap will bring him to his original starting position, but now he will be upside down.

In an analogous way Möbius found a method to turn inside out the modified harmonic series that is Riemann's zeta function. The Möbius inversion proceeds by first removing from consideration all integers that are divisible by the square of a prime. These are numbers like four, eight, and twelve, each of which is divisible by four (which is two squared), all multiples of nine (three squared) and twenty-five (five squared), and all larger squares too. This leaves only the terms of Riemann's zeta function which use natural numbers that are ***square-***

free, i.e., those positive integers which are not divisible by the square (or higher power) of any prime. Square-free numbers come in two flavors, characterized by their having either an odd or an even number of (necessarily distinct) prime factors. For example, two, three, and five are of the first kind (since they are prime and thus are the "product" of one prime factor), as are thirty and forty-two (each the product of three distinct prime factors). The numbers six, ten, and fifteen are all examples of the latter because each is a product of two distinct primes. The reciprocal of the zeta function is obtained by adding the reciprocal powers of square-free integers with an even number of prime factors, subtracting those with an odd number and ignoring all the rest.*

Another way to think of this is to consider each term in the infinite series that makes up the zeta function, and turn this into a new series by multiplying each term by either zero, one, or minus one, depending on whether it is meant to be ignored, added, or subtracted according to the considerations described above. The second row of the table in Figure 17 shows the first several Möbius values.

Encoded in this new assignment of numerators to the series of Riemann's zeta function is information about how its reciprocal grows, and properties of this reciprocal provide new insight into Riemann's zeta function and the Riemann hypothesis. Stieltjes realized that if he could prove that the accumulation of the plus ones and minus ones that constitute the numerators did not grow any more quickly than a multiple of the square root (i.e., that the sum of the first hundred values was roughly ten, and the sum of the first million was roughly one thousand, and so on . . .) then this would imply the truth of the Riemann hypothesis. The successive accumulation of the first several Möbius values are shown in the third row of Figure 17.

The primary record of Stieltjes's research on the Riemann hypothesis is to be found in his collected correspondence with Hermite. In these letters Stieltjes claims to have hit upon some "marvelous cancellations" allowing him to show that this seemingly unpredictable stream of plus and minus ones generated by Möbius inversion accumulates slowly enough to imply the Riemann hypothesis. Many of the

*So the reciprocal of Riemann's zeta function is itself an infinite series that begins: $1 - 1/2^s - 1/3^s - 1/5^s + 1/6^s - 1/7^s + 1/10^s \ldots$

Integers	2	3	4	5	6	7	8	9	10	11	12	13	...
Möbius values	−1	−1	0	−1	1	−1	0	0	1	−1	0	−1	...
Running total	−1	−2	−2	−3	−2	−3	−3	−3	−2	−3	−3	−4	...

Figure 17. The first twelve values assigned in the Möbius inversion formula, accompanied underneath by a running count of their successive accumulation. Integers that are divisible by the square of a prime are assigned zero, those that are square-free and have an even number of prime factors are assigned one, and those that are square-free and have an odd number of prime factors are assigned minus one. In particular, this means that individual primes, whose prime factorizations simply consist of themselves (i.e., a single prime), are assigned minus one by the inversion formula.

more than 400 letters they exchanged between 1882 and 1894 contain references to his proof, complaints of overwork, and promises of publication after simplification of the "too complicated" derivations.

In 1890—possibly as a way of pushing Stieltjes to publish his results—Hermite used his position as president of the French Academy of Sciences to organize a Grand Prize competition devoted to understanding the distribution of the primes, and in particular designed to engage mathematicians in the process of tidying up the unfinished business of Riemann's paper. Despite Hermite's repeated urgings, Stieltjes did not submit a manuscript. Several years later Stieltjes was dead, having neither recanted his claim nor produced a proof of the Riemann hypothesis. In the last of his letters to Hermite to mention the Riemann hypothesis, Stieltjes once again makes clear that he believes he has already settled the conjecture, and having done so, will abandon any further investigations of Riemann's explicit formula, saying, "I leave you with regret to follow all alone the path toward a country where all is unknown."

A chance that Stieltjes is right

Two years after Stieltjes's death, his ideas surfaced once again. The impetus seems to have been a new way of statistical thinking that had begun to pervade scientific thought. Physicists, economists, chemists, and mathematicians had begun to see the possibility of subjecting randomness to mathematical rigor. Chance was in the air, and the seemingly random zigzags of the Möbius function suggested to at least one mathematician a mathematical process that had only recently appeared on the scene, the ***random walk.***

The random walk is one of the most basic random processes, a mathematical model meant to account for phenomena like the seemingly unpredictable ups and downs of the stock market. It has its intellectual roots in the classic model of *repeated coin-tossing,* an imaginary game in which Peter and Paul, two infinitely patient and infinitely wealthy gamblers, repeatedly toss a coin and bet one dollar on each outcome. Peter always calls heads, leaving Paul to call tails, and with each toss the winner gains a dollar from the loser. We keep track of Peter's tally of wins over time. Perhaps it rises for a while as he gets lucky with a string of heads, only to plummet when he has an unlucky run of tails. Over time Peter's winnings wander unpredictably between positive and negative, between profit and loss.

This never-ending game has an infinite collection of possible outcomes, equally likely histories of the episodic waxing and waning of Peter's fortune. We have some intuition regarding the (unreachable) ultimate outcome of such a game: should Peter and Paul play forever, then in the infinite limit, at the temporal horizon when the ideal casino finally closes its doors, Peter and Paul step away from the table even. All that work (or play) is summarized in one simple fact: the most likely outcome is no change at all.

Along the way we can ask different kinds of questions about variation. What is the likelihood that Peter is ahead for some amount of time? How often is he ahead? The culmination of this sort of study is the ***central limit theorem,*** which among other things is the mathematical justification for the well-known bell curve. The central limit theorem predicts that after many plays of the game, we can expect that the majority of the time, the amount by which Peter either leads or lags behind Paul is within the square root of the number of plays. Thus, for example, after 10,000 plays, we can expect Peter's and Paul's winnings to differ by about 100; after 1 million plays we can expect them to differ by about 1,000.

The result of a computer simulation of four games of coin tossing

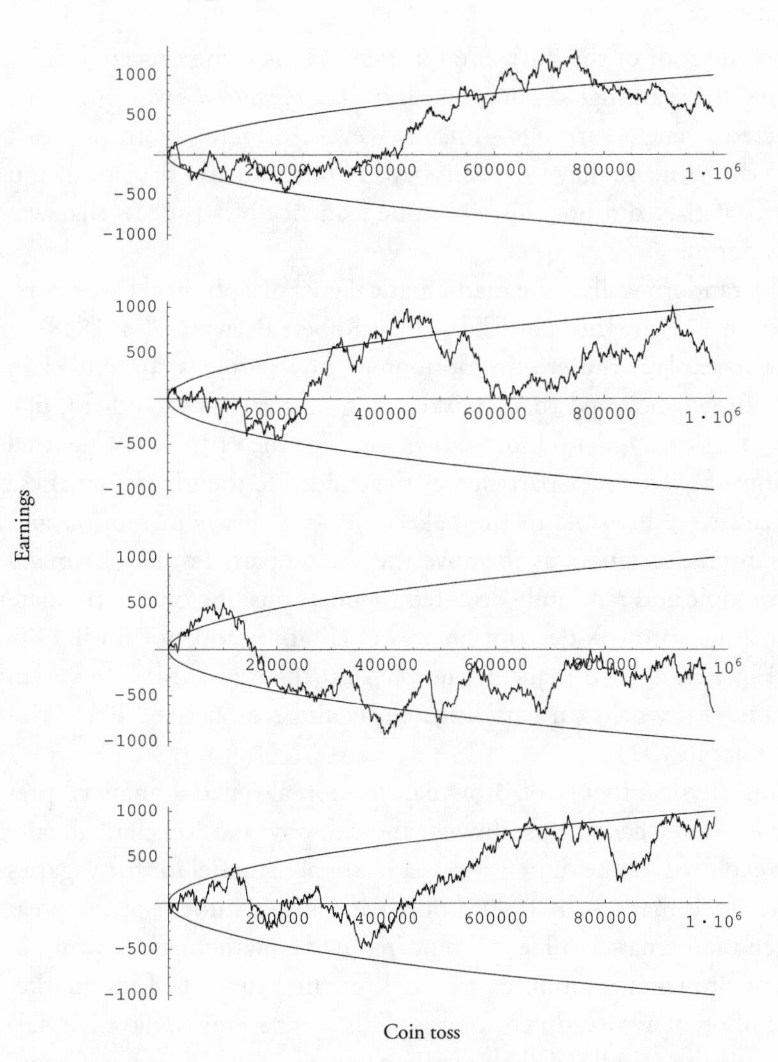

Figure 18. Four computer simulations of Peter's earnings in a game of 1 million tosses of a fair coin. The "envelope" around the jagged earnings line is a graph whose distance above and below the horizontal axis is equal to the square root of the horizontal distance.

between Peter and Paul in which the coin is tossed 1 million times is shown in Figure 18. In each graph we follow Peter's current earnings after each toss of the coin, so for example, in the first graph, after about 400,000 plays, Peter is even, but after 700,000 plays, he is up by more than $1,000. The "envelope" around the jagged earnings line is a graph whose distance above and below the horizontal axis is equal to the square root of the horizontal distance. This is the expected variation in this earnings report, one *standard deviation* away from the expected average earnings—breaking even, or zero. Notice that at times the earnings may vary even beyond one standard deviation, and indeed, if they did not this would be evidence of a process that was not random.

The random walk is the mathematical guts of a physical theory first voiced in 1827 by the Scottish botanist Robert Brown (1773–1858) in order to explain the erratic motions of tiny particles suspended in fluid. Brown believed that the skittering of particles to and fro, like water bugs on a quiet pond, was evidence for the existence of thermal motion of even tinier particles in the fluid. He theorized that these particles were invisible to the naked eye, but always in motion and substantial enough so as to move the visible particles, like so many microscopic and randomly oriented tugboats pushing on a particulate ocean liner. Brown's description was qualitative, and not until 1905 was Einstein able to make a true physical theory of this "Brownian motion," for which (in part) Einstein would receive the Nobel Prize in physics in 1921.

This physical theory of Brownian motion also had a financial predecessor. The chaotic ups, downs, and sideways motions had already been explored in one dimension as a reasonable model for the vagaries of the stock market. In 1900, Louis Bachelier, a student of the great French mathematician Henri Poincaré, used a mathematical formulation of Brownian motion as a model for the behavior of the market value of French government bonds. This work seems to have lain fallow until the middle of the twentieth century, when it was rediscovered by the statistician M. G. Kendall, and later popularized by Burton Malkiel in his best-selling book *A Random Walk Down Wall Street.* Brownian motion is also the intellectual engine powering the *Black-Scholes equation,* a technique which remade options pricing. As befits its up-and-down nature, the Black-Scholes equation was both the source of the 1997 Nobel Prize in economics for two of its discoverers, Robert Merton and Myron Scholes (codiscoverer Fischer Black died in 1995) and partially responsible for the founding and spectacular near collapse in 1998 of their Long Term Capital Management hedge fund.

Could the same mathematics that illuminated the random ups and

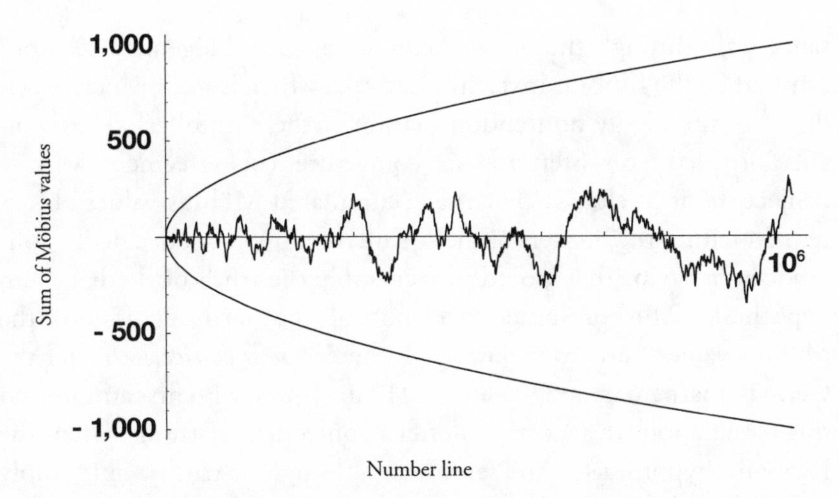

Figure 19. The first million values of the accumulated Möbius values, compared with the expected variation (one standard deviation) in tossing a fair coin 1 million times.

downs of the stock market, or the motion of dust on a pond, be able to explain the seemingly random sequence of plus and minus ones generated by the Möbius inversion formula?

In a paper published in 1897, a little-known Austrian mathematician, R. Daublebsky von Sterneck, reported the results of computing the successive accumulations of the first 150,000 values of the Möbius inversion formula. Von Sterneck asked if the Möbius inversion formula generates its plus and minus ones with equal likelihood—in his words, like a "die having plus one on three faces and minus one on the other three faces." After extensive numerical experiments he concluded that the variation is well below what would be expected from a random process.

The graph of the successive accumulation of the first million Möbius values is shown in Figure 19. The ups and downs of this theoretical coin tossing are well within the variation that would be expected if they were truly random. In comparison with the four graphs in Figure 18, each of which charts the progress of 1 million tosses of a fair coin, the ones and minus ones of the Möbius function seem to exhibit much less variation. Over such a long period of time, a truly random process would be expected to stray repeatedly beyond one standard deviation.

Von Sterneck was able to carry out his computations to such a distance only through the use of many complicated algebraic relations satisfied by the Möbius inversion formula, which is itself concrete evidence of the highly nonrandom nature of these numbers. These calculations led von Sterneck to conjecture (in agreement with a conjecture of Stieltjes's) that the accumulated Möbius values always remain within the borders of the square root graph—a decidedly nonrandom behavior that would in fact imply the truth of the Riemann hypothesis. This presumed behavior of the accumulation of the Möbius values came to be known as the ***Mertens conjecture,*** after the German mathematician F. Mertens (1840–1927), who first announced this speculation. In fact, the Mertens conjecture is stronger than the Riemann hypothesis in the sense that although its truth would imply the truth of the Riemann hypothesis, its falsity does not mean that the Riemann hypothesis is false. So, even though Stieltjes never did provide a proof of the Riemann hypothesis, he did at least leave another possible path for those pursuing it.

8 *A Prime Number Theorem, After All . . . and More*

Stieltjes's letters show that he believed that the Möbius function stayed within the square root lines, but he admitted he could not prove this. However, what he did claim (in his usual coquettishly apologetic manner) to have proved about the seesawing Möbius function is that it stayed within the slightly higher curve given by the three-fourths power. Had he in fact shown this, he would have "at least" proved the Prime Number Theorem. Unfortunately, this claim also must go down as a historical "what if."

Nevertheless, while Stieltjes's hemming and hawing echoed the random walks in his work, mathematics was not standing still. Real progress on the Riemann hypothesis was ahead, and these advances would shed new light on an evolving notion of numbers and their properties.

We have a winner!

After two years of waiting for a manuscript from Stieltjes, Hermite finally gave up on him. There were but two entries in his Grand Prize Riemann-related competition, and one was declared invalid for technical reasons having to do with the form of the manuscript. So a winner was declared: the young French mathematician Jacques Hadamard (1865–1963), just two years past receiving his doctorate. Although his entry was the only accepted submission, his win was definitely not a gimme.

Ironically, while working hard to wheedle out of Stieltjes a proof of the Riemann hypothesis, Hermite had also served as cheerleader for Hadamard's Grand Prize entry. The subject of Hadamard's doctoral dissertation, like Riemann's, was complex analysis, an extension of the calculus to complex numbers. As a member of Hadamard's thesis committee, Hermite had urged Hadamard to find some applications for the techniques developed in his dissertation. Inspired by Hermite, Hadamard reviewed the classic paper containing the Riemann hypothesis and found that he could use his own work to provide rigorous justification of a crucial but still unproved step in Riemann's derivation of the explicit formula for counting the primes. This achievement earned him the Grand Prize.

Hadamard's prizewinning memoir marked his first step toward understanding Riemann's work. His next step was more of a leap. In 1896, four years after winning the Grand Prize, Hadamard succeeded in pushing Riemann's ideas forward all the way to a proof of the Prime Number Theorem. At last, almost a hundred years after Legendre and Gauss had first conjectured a shape for the distribution of the primes, its truth was settled beyond the shadow of a doubt. The Prime Number Theorem was a theorem at last. Although the Riemann hypothesis remained unsettled, at least the basic topography of the primes was now mapped.

The proof of the Prime Number Theorem was a great moment for Hadamard, but in a surprising turn of events, it was a moment of glory that would have to be shared. While reviewing the galleys of his paper proving the Prime Number Theorem, Hadamard learned of the announcement of another proof of that theorem. This other proof, obtained by different means, was the work of a Belgian mathematician, Charles de la Vallée-Poussin (1866–1962).

Hadamard and de la Vallée-Poussin, who were almost exact contemporaries, make a strangely matched pair of mathematical twins. Hadamard, the son of an itinerant and irascible academic, would lead a life of professional highs and personal lows, the latter reflecting the tragedy and diaspora of European Jews of that generation. In contrast, de la Vallée-Poussin would spend almost his entire life close to the place of his birth, holding a professorship at his home university in Louvain for more than fifty years. Hadamard could claim relation (by marriage) to the religious-political outsider Alfred Dreyfus (of the infamous "Dreyfus affair"); de la Vallée-Poussin could trace his family—which included the French neoclassical artist Nicholas Poussin (1594–1665)—back to an entrenched nobility.

Edging toward the Riemann hypothesis

Beyond settling the Prime Number Theorem, the proofs provided by Hadamard and de la Vallée-Poussin are significant as the first rigorous steps toward confirmation of the Riemann hypothesis.

Riemann had already proved that any zeta zeros not on the real axis of the complex plane (i.e., any nontrivial zeta zeros) must in fact lie within the famous critical strip. This is the north-south highway bounded on one side by a breakdown lane—the imaginary axis—and on the other side by the line of complex numbers with a real part equal to one. According to the Riemann hypothesis, the zeros lie only on the highway's divider, the critical line in which all numbers have a real part equal to one-half. Hadamard's and de la Vallée-Poussin's proofs of the Prime Number Theorem both started the necessary process of containment, or conversely, the process of showing that either side of the critical strip is free of zeta zeros.

Even at the moment of this historic mathematical achievement by Hadamard and de la Vallée-Poussin, the shadow of Riemann loomed large, as did Stieltjes's famous claim. Hadamard prefaced his accomplishment with a nod to Stieltjes, writing that although Stieltjes had "proved" that the zeros of zeta function were in accordance with Riemann's assertion, Hadamard himself needed to prove "only" that the borders of the critical strip were zero-free in order to prove the Prime Number Theorem. Stieltjes or no Stieltjes, the walls were slowly beginning to close in on the critical line. But the power and importance of a conjecture like the Riemann hypothesis lie not simply in its challenge to the great minds of the day, but also in the progress across mathematics that is achieved as a result of its pursuit. Indeed, in the course of mining Riemann's paper to settle the Prime Number Theorem, Hadamard and de la Vallée-Poussin were able to push forward other prime frontiers.

Periodic tables of the primes

If the primes are the atoms of arithmetic, then the proof of the Prime Number Theorem marks the beginning of an understanding of their organization, providing a broad picture of their overarching distribution in the world of number. But the initial progress made by Hadamard and de la Vallée-Poussin toward a description of the zeta zeros would also have implications for a classification scheme for the primes accomplished by means of arithmetic progressions.

As we saw, Dirichlet was the first to prove that in almost any arithmetic progression an infinity of primes would be encountered. But the description of such a progression as the result of a hop, skip, and jump along the natural numbers can be restated as a listing of those numbers characterized as having the same remainder when divided by a given natural number, which is the step-size of that progression. For example, when we vault through the integers, taking them two at a time, the every-other nature of this walk lands us on either only even integers or only odd integers, depending upon where we start. This is the colloquial characterization of, respectively, the numbers that produce either a remainder of zero upon division by two or a remainder of one. With this comes a first distinction among primes, into the families of even and odd, the former containing only the prime two, and the latter all the other primes.

So, evenness and oddness are properties revealed by performing division by two. We arrive at other characterizations by considering division by other numbers. The three distinct and disjointed arithmetic progressions that we obtain by bounding along by three provide a trichotomy of the integers according to the remainder obtained in division by three. In this case, only the prime three gives a remainder of zero, while the other primes are scattered among the lists of those numbers giving remainders of one or two.

When Mendeleev performed an analogous categorization with the physical elements, grouping them according to physical properties such as reactivity, texture, and melting point, he arrived at an organization that turned into the periodic table. Its rows reflect an octave structure of the elements (something first observed by a British chemist, John Newlands) by distinguishing properties that seemed to repeat every eight elements. The eight distinct types of elements discovered (the "groups") include the inert or noble gases, halogens, alkalies, and alkaline earth metals. In a similar fashion we can organize the primes according to the effects of division by a fixed number, and so

Primes with remainder of 1 after division by 4	Primes with remainder of 3 after division by 4
5	3
13	7
17	11
. . .	. . .

Figure 20. "Divide-by-Four Periodic Table" of the primes, which organizes the primes according to their remainder after division by four. The first column shows the primes with remainder one, the second those with remainder three. There is no column corresponding to zero, as no prime has remainder of zero after division by four (that would be equivalent to being divisible by four). Only the prime two has a remainder of two after division by four, so we omit that column.

are able to construct an infinity of numerical periodic tables in which the primes are organized into columns that reflect a common behavior among primes when subjected to the knife of division. Figure 20 shows one such periodic table formed by using division by four.

Dirichlet's invention of the L-series to prove an infinitude of the primes in almost all arithmetic progressions reveals that generally the columns of these arithmetic periodic tables have infinite length. But even more, just as the Prime Number Theorem gives a more textured version of Euclid's infinity of primes, so too we can ask for a more detailed understanding of the infinite columns in our arithmetic prime tables. At what rate do their columns grow? In particular, are the primes *equidistributed* by remainder? That is, do the columns of the primal periodic tables grow at the same rate?

This is the second question that Hadamard and de la Vallée-Poussin were able to address with their newly discovered analytic techniques. In this they again followed Riemann's lead. Riemann extended the harmonic series to the complex realm in order to create the zeta function, thereby fashioning the key to proving the Prime Number Theorem. Hadamard and de la Vallée-Poussin then extended Dirichlet's L-series to the complex realm in order to study the equidistribution of the primes according to remainders.

Dirichlet L-series are variations of Riemann's zeta function and thus share the property that their behavior can be understood by the location of the zeros in the complex plane. In particular, the zeros of the L-series are just like the zeros of the zeta function: it is not difficult to prove that in this case too the zeros must fall in the critical strip, and it is conjectured that they too lie on the critical line. This is the form of the *Riemann hypothesis for L-series.*

As in the case of Riemann's zeta function, Hadamard and de la Vallée-Poussin showed that even for these complexified Dirichlet L-series, the borders of the critical strip remain zero-free. This fact implies the equidistribution of the primes. So, for example, our periodic table above, which begins to organize the primes according to their remainders upon division by four, will in the long run have the look of a rectangle. In the long run, up to any given point, a prime is approximately equally as likely to produce a remainder of one as three.

FIRST WE ORGANIZE, THEN WE EXAMINE: GAUSSIAN INTEGERS AND GAUSSIAN PRIMES

But things do not end here. Mathematics, like any intellectual pursuit, is an ongoing process of organization and examination. Just as the evolution of fundamental units of matter from elements to atoms initiated new inquiries into the nature of the world, the additional possibilities implicit in the extra degree of freedom of complex numbers made possible a new and more general notion of prime number. These explorations were led by Riemann's contemporaries and successors. Once again, Riemann's zeta function will be a tool critical to an organization of the notion of prime in what is now an evolving understanding of number extended from the real line to the complex plane.

First we must ask, What does *integer* mean in the two-dimensional world of complex numbers? Surely it should be something more than the regularly spaced dots that fall along the single line of the complex plane that is the real axis. In the complex world that has both length and width, how do we make more general the notion of integers?

Let's use our imagination—or at least imaginary numbers. A natural first idea is to simply mimic on the imaginary axis the evenly

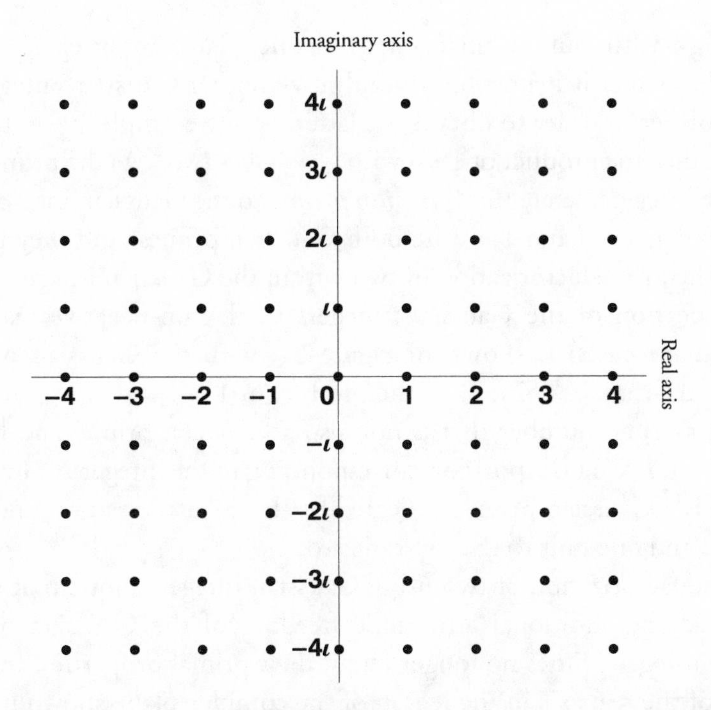

Figure 21. Partial picture of the latticework of the Gaussian integers in the complex plane. They are the complex numbers given by all possible combinations of integers and integer multiples of i, so that they give this regular array of points.

spaced signposts of the real axis that mark the familiar integers. Thus, we now include their imaginary counterparts, paired with Kronecker's God-given natural numbers and their man-made negatives: . . . −2, −1, 0, 1, 2, . . . and −2*i*, −*i*, *i*, 2*i*, and so on. But if these "new" integers are to include both the real and the imaginary, then their mixtures must be allowed as well, so that *integer* now takes on the meaning of all complex numbers obtained by sums of all possible integer multiples of 1 and *i*.

Graphically, we extend the regularly spaced steps that produce the integers on the number line to rectilinear motions in two dimensions. Instead of only steps to the east or west, we permit north-south motion as well, creating a checkerboard, or latticework, of points in the complex plane defined as those points with integer coordinates. These are the ***Gaussian integers.***

What of divisibility in this larger world? What are the primes? In an analogy with our familiar integers, one Gaussian integer *divides* another when it is possible to multiply another Gaussian integer by the former in order to obtain the latter. So for example $1 + i$ divides two, since the product of $1 - i$ with $1 + i$ gives two.* As the primes are to the integers, so are the ***Gaussian primes*** to the Gaussian integers. In particular, $1 - i$ and $1 + i$ are both Gaussian primes, and they represent the prime factorization of two within the Gaussian integers.

A portion of the Gaussian integers (with non-negative real and imaginary parts) is shown in Figure 22, with the Gaussian primes marked by an "x." Since 2 is the product of $1 + i$ with $1 - i$, it is an integer prime number that is not also a Gaussian prime, and hence there is no "x" at the position corresponding to the integer 2. However 1 + i is a Gaussian prime, indicated by the "x" at a position one unit above and one unit to the right of zero.

The factorization of two using Gaussian integers shows that when allowed the additional arithmetic freedom of the Gaussian world, some integer primes no longer retain their primal properties. Inspection of the real axis in the region of the complex plane shown in Figure 22 indicates that other integers which lose their primal nature in this larger world include five, thirteen, and seventeen, whereas primes such as three, seven, eleven, and nineteen all retain their primality.

Those integer primes that are not also Gaussian primes (e.g., five and thirteen) have in common the property that when divided by four, they give a remainder of one; those that remain prime (e.g., three and seven), and are hence both integer primes and Gaussian primes, produce a remainder of three when divided by four. So division by four becomes an acid test for the primes, providing a general rule through which we can distinguish between those primes that are *inert* (i.e., are both integer primes and Gaussian primes) and those which *split* (i.e., can be factored) in the Gaussian realm, thereby giving a new characterization of the columns in the "divide-by-four periodic table" of Figure 20.

Further arithmetic investigations reveal that the primes which split may be represented as the sum of squares of two integers: five as the sum of one and four, thirteen as the sum of four and nine, seventeen

*$(1 + i) \times (1 - i) = 1 + i - i - i^2 = 1 - (-1) = 2$

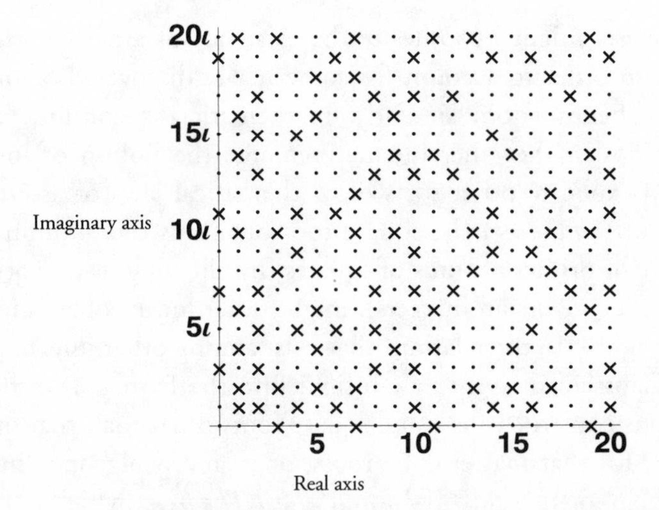

Figure 22. "X" marks the spots of the Gaussian primes among the dotted array of the Gaussian integers.

as the sum of sixteen and one, and so on. These are the primes that Pythagoras would love, those whose square roots may be realized as the length of the hypotenuse of a right triangle with integer-length legs.*

Thus at the heart of the phenomenology by which we created the divide-by-four periodic table of the primes, a classification according to remainders of one or three, we find properties that reveal themselves only when we start to investigate the properties of the Gaussian integers, the effect of positing and using the square root of minus one with the integers. Just as a hypothesized electron would begin to account for the structure of the organization of the basic elements of nature that is the periodic table, the imaginary number *i* provides the key to an analogous organization of the fundamental building blocks of the natural numbers. More than simply a surprise weapon in one of Cardano's mathematical duels, this imaginary number gives us another piece of ammunition in the never-ending game of "countably many questions" that mathematicians play with number, and physical scientists play with the world.

*Recall the Pythagorean theorem: the sum of the squares of the lengths of the legs of a right triangle is equal to the square of the length of the hypotenuse.

Physicists might hypothesize the existence of a new fundamental particle in order to account for some newly discovered or imagined natural phenomenon; similarly, mathematicians continue along a road in which they incrementally enlarge the notion of integer in order to explicate other sorts of mathematical laws or assertions. If tossing in (*adjoining*) the square root of minus one illuminates the structure of primes as sums of squares (by allowing us to work in the realm of the Gaussian integers), might other means of augmentation yield insight into the relations that exist among other integer powers? If the Gaussian integers gave insight into the Pythagorean theorem, analogous constructions using square roots other than that of minus one—or for that matter cube roots, or even any algebraic integer—might help in the study of ***Fermat's Last Theorem,*** which is a natural generalization of Pythagoras. This assertion (proved in 1994 by Andrew Wiles of Princeton University, some 350 years after it was first made) is that the cube of any integer is never the sum of two nonzero cubes, the fourth power of any integer is never the sum of two other nonzero fourth powers, and so on.*

But in using these new number systems we must be careful. Although the Gaussian integers retain many of the usual properties of the everyday integers (sometimes called *rational integers*), it can happen that in other superficially similar worlds, natural mathematical law seems to fall apart. For example, suppose we try, using the Gaussian integer system as a model, a number system in which instead of incorporating among the integers the square root of minus one, we use the square root of minus five. That is, we create a numerical world in which integers include numbers like $1 + \sqrt{-5}$, $-1 + \sqrt{-5}$, $2\sqrt{-5}$, or $7 + 24\sqrt{-5}$, i.e., sums of integers and integer multiples of root five, akin to the Gaussian integers. Unfortunately, in this world, integers "misbehave." Although here there is still a notion of prime, defined as a number which cannot be written as the product of two smaller numbers, it is possible that a number can have two different expressions as a product of primes. In other words, in this root-minus-five world (as opposed to the root-minus-one world of the Gaussian inte-

**Nonzero* is an important specification. Notice that if you let yourself use zero, then you can write any cube as a sum of two cubes. For example, consider 8 (which is 2^3). We can write: $2^3 + 0^3 = 2^3$. The same trick works for any power and any number.

gers) prime factorizations are no longer uniquely determined. In these situations we find ourselves in the same position as the physicist who needs to account for the nature of light as both an irreducible particle and a wave. How to sustain the overarching principles of arithmetic in light of this new information?

For the physicist, the consolidation of wave and particle was made possible by the invention of quantum mechanics. Here, the seemingly irreconcilable fact of disparate prime factorizations was resolved by the invention of the *ideal number,* or *ideal.*

In looking for a way to use these Gaussian-like ideas for proving Fermat's Last Theorem, Riemann's contemporary Eduard Kummer (1810–1893) came up with the idea of ideal numbers, thereby creating ***algebraic number*** theory. This is the counterpart of analytic number theory, the subject area of our Riemannian story. Whereas analytic number theory takes on questions of asymptotics and distributions of the properties of the integers (i.e., examines the properties of numbers in the large), algebraic number theory instead studies more the laws that dictate individual interactions between numbers. Algebraic number theory asks questions such as what kinds of integers can be represented as the sum of two integer squares (i.e., are solutions to the Pythagorean theorem), or two integer cubes, or other powers—in other words, it considers how algebraic relations (addition and multiplication) dictate the ways in which integers can combine to make other integers. This is akin to the manner in which laws of physics and geometry constrain the particular sorts of molecules into which individual atoms can aggregate.

Whereas Kummer is credited with the origins of algebraic number theory, Richard Dedekind (see Chapter 6), Riemann's friend and biographer, was the mathematician responsible for its development and axiomatic investigation. Like Riemann, Dedekind was a student of Gauss's (his last). Dedekind received his "habilitation" degree almost simultaneously with Riemann, and also benefited from a close friendship with Dirichlet. However, Dedekind, who did not have Riemann's financial worries, seems to have opted out early from the academic rat race. He never married; he was cared for by one of his four sisters; and he spent almost his entire professional life teaching at a technical high school in Berlin.

Within this quiet and orderly world Dedekind reinvented the

meaning of number in all its variegated forms. He invented the notion of ideal number to recover prime factorization in number fields, but he also took seriously the question of just what a real number is. What would it mean to have "root two" pounds of rutabaga? What could it mean to have a never-ending decimal expansion of π pies? Taking the rational numbers as given, Dedekind came up with the idea of defining real numbers in terms of the way they partition the rational numbers into one collection consisting of all those that are less than a given number and another consisting of all those that are at least as large as the given number. This definition in terms of *Dedekind cuts* is still one of the standard constructions of the real numbers.

To begin organizing his enlarged world of ideal numbers, Dedekind adapted Riemann's techniques, creating the *Dedekind zeta function.* As with its predecessor, its properties depend on its zeros, again restricted to the critical strip, and again hypothesized to lie on the critical line. Any information about the zeta zeros will have implications for the Dedekind zeta function, and for the Dirichlet zeta function. To understand the zeta zeros and the familiar prime numbers will be to understand the accumulation of primes in all their many incarnations and elaborations.

THE RISING OF THE RIEMANN HYPOTHESIS

The tools introduced by Riemann in his brief memoir paved the way to a proof of the Prime Number Theorem, which had been one of the great unsolved problems of the day. Even more, the basic ideas and techniques in Riemann's short paper had also pointed the way to a deeper understanding of numbers and the essence of primality. All these achievements required only that the border of the critical strip remain zero-free. But all this understanding had not required the resolution of the Riemann hypothesis.

Still, as is so often the case in science, and in life, the more you know, the more you realize what remains to be learned. We are greedy; we are inquisitive. The story of science is, in part, an infinite tape loop of the tale of Icarus and Daedalus, one generation's achievement serving as the means and motivation for the next generation's aspirations.

Grandparents trudge to the corner store, parents learn to fly over the oceans, and children dream of reaching the moon. The sequencing of the genome was once a goal; now it is but a stepping-stone to the understanding of man, disease, and the nature of life itself.

Mathematics is no different. Now that we know the coarse behavior of the primes, we might ask, can we do better? As it turns out, only if we can settle the Riemann hypothesis.

9 *Good, but Not Good Enough*

IN THE PROOF of the Prime Number Theorem we have a beautiful, simple, elegant statement: that at the outer reaches of the number line, the number of primes less than a given number is essentially the ratio of this number to its logarithm. This beauty is poetic in its haikulike economy of language. It uses only the most familiar of mathematical objects, numbers and their logarithms, the former modulated by the reciprocal of the latter. The former describes a growth of the count as we consciously count, returning each number at its face value; the latter, the scale of the logarithm, the unit of the decibel or Richter scale, is the count of the senses. These are the twinned functions of perception and sensation. That together they should give us so much information about the primes is as close to poetry as mathematics might ever hope to come.

But we know that we've been sloppy, and it nettles us. The devil is in the details, and it is the details that we've left behind as we race off to infinity. Just how much were we compromising by looking at the primes from beyond the beyond? From the stratosphere we see the big blue marble that is earth, but we miss the beautiful variation of the local landscape.

It could very well be that the way the count of the primes approaches the truth of the Prime Number Theorem is less a determined march toward equality than a wild, meandering, and loopy drunkard's walk, careening above and below the smooth curve representing the ever distant, infinite, unreachable asymptotic epigrammatic formula first posited by Gauss.

The work of Riemann, Hadamard, and others showed that the asymptotic behavior of the count of the primes is determined by the zeta zeros, their positions acting as an overarching influence on its distant activities. So the zeros of Riemann's zeta function will help us to understand the pace of the walk to the limit of the distribution of primes. This was the quest begun by de la Vallée-Poussin, who would reveal the true import of the Riemann hypothesis.

To err is Riemann

Nicholas Poussin is known for his landscapes inspired by mythology, a painterly merging of the Platonic with the real. His relation, de la Vallée-Poussin, took up an analogous number-theoretic study, attempting to bridge the actual and the unattainable by investigating the difference between the limiting behavior of the primes that is the Prime Number Theorem and the real-life number-by-number measurement that is their exact count.

De la Vallée-Poussin closed the nineteenth century by finally proving that the location of the zeros of the zeta function is directly related to understanding the accuracy of the approximation by means of the Prime Number Theorem, or, more specifically, by Gauss's estimates using the logarithmic integral. Riemann's explicit formula had provided a count of the primes in terms of the zeta zeros, so it was almost surely no great surprise that their location in the complex plane, i.e., their values, would influence the approximation. Since Riemann's exact formula is an infinite sum of primal waves, as each zero is tossed into the mix, and another wave is brought crashing onto the jagged prime coast, the approximation becomes closer and closer to the exact truth. The *error term* measures the difference between the exact count and the approximation, and we take up its discussion here.

De la Vallée-Poussin made progress in two directions related to the Riemann hypothesis. First, he edged slightly closer to the confirmation of the Riemann hypothesis, tightening ever so slightly the net around the zeros of Riemann's zeta function, a net which currently lay just within the borders of the critical strip. De la Vallée-Poussin proved that the net could be drawn tighter toward the bottom of the strip, while blousing out to its borders as we move farther up the critical strip. A bit more precisely, he showed that the region within the critical strip known to be free of zeta zeros starts out with its borders definitively within those of the critical strip at the bottom, but then gradually moves out toward the edges, although never quite reaching them.

With this small improvement in the determination of the precise position of the nontrivial zeta zeros, de la Vallée-Poussin was then able to begin to understand the degree to which Gauss's estimate for the accumulation of the primes (which in the large gives the estimate of the Prime Number Theorem) missed the exact count. In this way, de la Vallée-Poussin was able to show that our understanding of the degree to which the coarse estimates of the prime accumulation miss the details is directly related to the locations of the zeros. He was also able to use these techniques to prove once and for all that Gauss's approximation of the count of the primes via the logarithmic integral, while asymptotically the same as that of Legendre or Encke, was in fact (as claimed by Gauss) a much better approximation (see Figure 7). Chalk up another one for old C.F.G.

Nicholas Poussin once wrote that his work reflects the fact that "I am forced by my nature toward the orderly." In beginning to treat the unruliness that is the detail between the exact and the asymptotic, de la Vallée-Poussin surely would have made his ancestor proud.

THE COMING CENTURY

Von Sterneck's efforts and the analytic results of de la Vallée-Poussin provided a hint of what the twentieth century would bring to the pursuit of the Riemann hypothesis: a marshaling of intellectual forces from probability and physics, coordinated with the insights provided by computation, all turned to the purpose of settling a conjecture that had begun to assume a central role in the progress of modern mathemats. Riemann's zeta function and his related hypothesis had become a template for an infinite collection of zeta functions, which were the key to understanding the fine structure of an evolving world of number, one that spread from the ideal primes of Euclid to the prime ideals of Dedekind. Zeta functions were the tools that would quantify the distance between approximation and truth, encapsulated in a *gen-*

eralized Riemann hypothesis. This tangle of conjectures inspired by the original Riemann hypothesis predicted that every one of the zeta functions, whether due to Riemann, Dirichlet, or Dedekind, vanished on a crisscross of zeros in the complex plane: trivial and known zeros on the horizontal, nontrivial and hypothesized zeros on the vertical.

But all things redound back to the original Riemann hypothesis. As the twentieth century began, the Riemann hypothesis remained a conjecture—plausible, but still neither true nor untrue. It was now time to launch a concerted intellectual assault.

A CALL TO ARMS

The excitement felt at the turn of the twenty-first century was a replay of many of the same emotions that had been experienced at the dawn of the twentieth. The buzz of the latest technology revolution is an echo of that initiated by an industrial revolution of the early 1900s. Then, as now, there was a spirit of progress in the air, a zeitgeist thick with dreams of exploration.

The early 1900s were a can-do time in which nothing seemed beyond reach. In a garage in Ohio the Wright brothers were designing the first airplane. Henry Ford was making the Model T, and it was being mass-produced through a modern technological innovation, the assembly line. Geographic boundaries were being surmounted as one intrepid soul after another sought to make the world small through conquest and travel. The North Pole and South Pole were soon to be reached. Nothing was beyond the grasp of man.

This same spirit of optimism and arrogance pervaded mathematics as well. So it was that on the threshold of the twentieth century, on August 6, 1900, in a lecture hall in Paris, 250 of the world's leading mathematicians settled into their seats to hear a lecture entitled "Mathematical Problems," delivered by one of their leading lights, Professor David Hilbert (1862–1943) of the University of Göttingen. This lecture, addressed to the Second International Congress of Mathematicians (ICM), would guide much of the research agenda for the mathematical world for the next century and beyond. Hilbert, who would lead mathematics into the twentieth century, had been born near the bridges of Königsberg that led Euler to develop topology and graph theory. After a meteoric rise through German academia, Hilbert found himself two generations removed from Gauss as professor of mathematics at the University of Göttingen, having assumed that position almost exactly 100 years to the day after Gauss had enrolled there. Hilbert followed in the footsteps of these mathematical giants in both person and spirit, as the depth and breadth of his mathematical abilities and interests made him their natural intellectual heir.

Hilbert first made his name in the solution of an age-old problem in *invariant theory,* a subject covering the interplay between coordinate systems—first developed by the philosopher and mathematician René Descartes (1596–1650)—and the manner in which they represent geometric objects. Descartes's mathematical creations, such as the Cartesian plane and Cartesian coordinates, translated geometric statements from the formally poetic prose of Euclid to the language of algebra. Points, lines, planes, circles, and triangles are no longer described in Zen-like koans ("a point is a dot of no spatial extent") that seem to do little more than confirm our inability to bring the Platonic world to earth. Instead, they are given an objective description in terms of a certitude of number and equation. What we lose in poetry we gain in precision and insight. Above and below become lesser and greater; coincidence becomes the simultaneous solution of several sets of equations. No longer is this a geometry whose truths are clarified solely by pure thought or capable draftsmanship. Calculation becomes the chief handservant to deduction. Enter algebra, the ability to manipulate the language according to a collection of agreed-upon rules and regulations meant to define the legal combinations of our symbolic language. In this we have the beginnings of the subject that is now called ***algebraic geometry.***

Any geometric object can have many different representations in this fashion (i.e., the same collection of points in space can be described in a variety of ways as the intersection of lines or curves), and conversely, a given way of representing space (i.e., a given coordinate system) may not provide the optimal description for a particular sort of configuration. Those aspects of a geometric configuration that remain unchanged over its various algebraic descriptions are called its *invariants.* Understanding the invariants of a system is important today in various aspects of computer-automated and computer-aided design.

Hilbert's work in invariant theory had many of the hallmarks of his later research. He discovered a basic procedure by which all invariants could be produced from an initial collection of basic invariants in a formal, almost mechanical manner, as if by a recipe, just as the cars on a production line are composed of their basic automotive elements. These were the days before the invention of the computer, yet Hilbert's work seems to foreshadow the computational revolution that would arrive in the middle of the twentieth century. In this sense his work is very modern.

Throughout his career, he would often take a programmatic approach, in which a field or subject is reimagined as the natural and logical production of facts from a few basic initial truths. In particular, turning away from invariant theory, he had quickly shifted fields to concentrate on rebuilding algebraic number theory, the study of how the integers fit together as the solutions to certain kinds of equations. In this work Hilbert unified the work of his predecessors, finding a coherent general formulation of the notion of "prime" which included Euclid's prime numbers as well as the prime "ideal" numbers of Dedekind and Kummer.

From number Hilbert once again executed an abrupt intellectual about-face and turned to the study of geometry. In his landmark treatise, *Foundations of Geometry,* he revisited the axioms of Euclid, rewriting them so as to include under one logical framework both the classical truths of Euclid and the newly discovered non-Euclidean geometries.

Thus, piece by piece, Hilbert seemed bent on re-creating mathematics from the ground up, rebuilding foundations where necessary, giving each subject a modern makeover so that it might be up to the challenges of the new century.

And in the Eighth Position, Playing Central Problem . . .

Hilbert's lecture to the ICM, "Mathematical Problems," delivered in the shadow of the Eiffel Tower, would form a scaffold upon which much of this future mathematics would be hung. In his lecture,

Hilbert presented a list of twenty-three problems or mathematical challenges, twenty-three swords waiting to be pulled from the stone. Together they would serve as a beacon to guide the development of much of modern mathematics, calling to the strongest and wisest. To make significant progress on, or to solve, a "Hilbert problem" is a guarantee of a place in the mathematical pantheon.

But these problems were more than mere challenges for the sake of challenge. Hilbert had suggested that these mountains be climbed not simply because they were there, but rather because he believed that the ensuing intellectual expeditions stood a good chance of trekking through new and never-before-seen fertile areas of research, journeys that seemed likely to arrive at a view of the mathematical landscape capable of reshaping the way in which mathematicians and scientists approached the world.

How did Hilbert choose these twenty-three problems from among the universe of unanswered mathematical questions? How to decide in advance which problems are deep? Which problems are interesting? Which problems are likely to reveal new and exciting connections among seemingly disparate subjects?

Hilbert was guided by considerations that to this day provide an important set of criteria by which any mathematical problem may be judged.

Presentation is important. A good problem, like a good story or a good meal, is immediately attractive. So, first, a good problem must lend itself to a clear and simple statement, for as Hilbert says, "What is clear and easily comprehended attracts; the complicated repels." Furthermore, a good problem should pose enough of a challenge that we are intrigued, yet not so much that we quickly grow frustrated.

Not only should the statement of the problem be a model of clarity, but the solution should be as well. Hilbert credits an anonymous French mathematician with the remark, "A mathematical theory is not to be considered complete until you have made it so clear that you can explain it to the first man whom you meet on the street." It is best to present a solution as a finite sequence of connected logical steps, the totality of which rests upon a few clear and logically consistent axioms. This is a mechanistic view of man and mathematician, line workers fitting together subassemblies of axioms, which are to be combined in a succession of logical constructions, in order to produce beautiful, economical, and supremely reliable expressions of thought; or instead, like a modern-day computer, turning the crank on the input of a clearly stated problem, in order to produce a solution in finite time.

Using these desiderata, Hilbert made his selections. To look at the list of "Hilbert problems" is to see a snapshot of modern mathematics and its evolution: the mysteries and paradoxes of set theory and the infinite introduced by Cantor; questions which aim to shed light on the logical foundations of number theory, geometry, and the mathematics of physics; questions of symmetry and dynamics that are critical today in applications such as planning the motion of robots; the study of partial differential equations that model climate change and fluid flow; hints of the future arrival of hybridized subjects like modern *algebraic geometry* and *algebraic topology.*

There, right in the middle of things at number eight, is the Riemann hypothesis. Of this, Hilbert declares that "it still remains to prove the correctness of an exceedingly important statement of Riemann." It is worth noting that Fermat's Last Theorem did not make the cut.

Near the close of the introduction to his list of problems, Hilbert issued a call to arms: "The conviction of solvability of every mathematical problem is a powerful incentive to the worker. We hear within us the perpetual call: There is the problem. Seek its solution. You can find it by pure reason, for in mathematics there is no *ignoramibus.*" And with that, the "workers" returned to their stations, axioms in one hand, pure reason in the other, recommitted to the production of mathematics, and in particular to settling the Riemann hypothesis.

An error term worth the effort

Should we find that the Riemann hypothesis is true, then the zeros of Riemann's zeta function are beautifully and symmetrically arranged, some on the real axis and others on the critical line, dotting these two axes like irregular landing lights lining the center of two perpendicular airstrips stretching out to infinity. In this case the zeta zero marks on this infinite X will be the source of a deep understanding of the

Figure 23. The von Koch snowflake.

distribution of the primes. If this is how the zeroes are organized, then the error term, the running tally that tracks the difference between the exact count of the number of primes and the asymptotic estimate of the Prime Number Theorem, will itself have a wonderfully simple asymptotic behavior, almost as simple as that which describes the primes.

In 1901 the Swedish mathematician Niels von Koch (1870–1924) took aim at the jagged line that details the primes. Complicated curves were his meat and potatoes. He remains famous for the construction of the "Von Koch snowflake," one of the most famous and recognizable fractal pictures.

Von Koch showed that proving the Riemann hypothesis implies that asymptotically, the absolute error (or *difference*) in using Gauss's logarithmic integral to estimate the number of primes is on the order of the logarithm times the square root of the stopping place. Or, in short, that the curve that is lurking beneath the jagged graph of Figure 7 is like the curve that charts the growth of the mathematical function that is the logarithm times the square root. Rather than look at the absolute error, we can also consider the *relative* or *percentage error* which measures the ratio of the difference between Gauss's estimate and the true count of the primes to Gauss's estimate. Von Koch's work implies that the curve charting the growth of relative error looks like

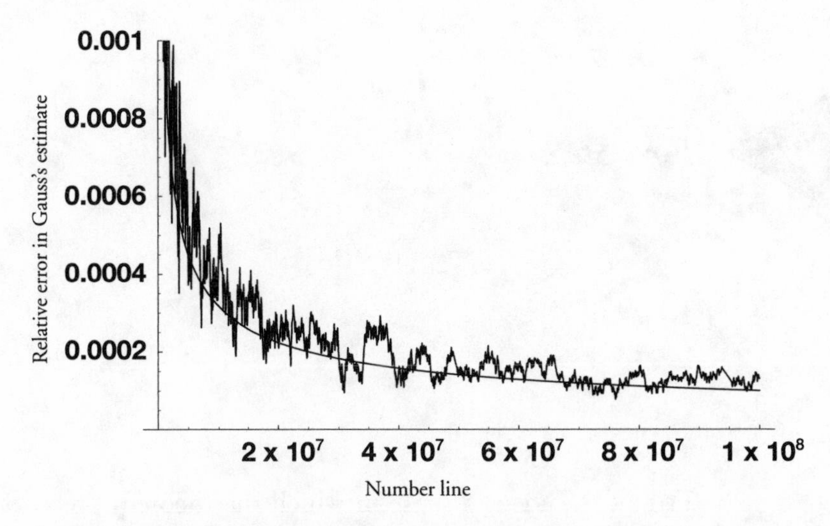

Figure 24. Relative error in using Gauss's estimate for the count of the primes. Smooth line is the graph of the reciprocal square root.

the reciprocal of the square root of the stopping place. Figure 24 shows the graph of the relative error with the graph of the reciprocal of the square root. This is an error term worth achieving, a characterization of difference which we can understand. Over the next seventy years, the mathematics factory would produce a wealth of evidence in its favor.

10 *First Steps*

THERE IS OFTEN a sense in which the theorem precedes the proof. It's a gut feeling, a sense that something "must be right," along with a small sheaf of notes, or maybe even just a few scratchings on a napkin. Within the collective gut of the mathematical community was a feeling that the Riemann hypothesis had to be true. Nevertheless, for all the feeling and poetry, it's also nice to have some hard data. Talk is cheap. Show me the zeros!

The gleam of a computational future that shimmered throughout Hilbert's address was reflected in the successive zeta-related work directed at computing zeros of Riemann's zeta function. Mathematicians (and later, a new breed of scientist, the computer scientists) turned their talents toward finding numerical evidence in support of the Riemann hypothesis, or perchance a quick path to mathematical immortality by discovering a zero that contradicted Riemann's assertion: a zeta zero off the critical line. Riemann's notes reveal that, working with just pencil and paper, he had managed to show that the first several zeros satisfied his conjecture, and indeed it was almost surely these first few checks that gave him the confidence to voice his conjecture.

The first published calculation of zeros in the critical strip is due to J. Gram (1850–1916). Gram, who was Danish, was an exceedingly interesting character. "By day" he was a prominent insurance executive who eventually rose to the position of chairman of the Danish Insurance Council, but Gram had a second life as an accomplished working mathematician. He wrote seminal papers in forestry management, and today is still a household name (in mathematical households) for his contributions to one of the most important computational techniques in the area of *applied linear algebra,* a subject at the heart of many of the automated procedures that produce the purchasing "recommendations" filling our e-mail in-boxes, a fitting legacy for an executive presiding over insurance salesmen. During his lifetime he was the recipient of several prizes for his mathematical research, including a gold medal for a paper related to the distribution of the primes.

In 1903 Gram published a list of the first fifteen zeta zeros, each computed to a precision of six decimal places; all fifteen (of course) were found to lie on the critical line. These computations were based on a particular expression for Riemann's zeta function called the *integral form* and used a relatively elementary calculus-based technique to calculate the zeros. By 1925, this approach had been used to confirm that the Riemann hypothesis was true for all zeros in the critical strip up to a height (i.e., with an imaginary part) of 300.

These initial computational confirmations make up half of what had become a two-pronged assault on the Riemann hypothesis. The "solution-seekers," the mathematical foot soldiers who hunt for zeta zeros, were matched by "solution-squeezers," the mathematicians who instead sought to eliminate large regions of the critical strip as possible sources of zeta zeros. They would squeeze the tube of provable potential zeta-zero locations, centered about the critical line, aiming for the day when only the critical line would remain.

Along these lines, the pioneers had been Hadamard and de la Vallée-Poussin, who in their respective proofs of the Prime Number Theorem had shown that none of the nontrivial zeta zeros had a real part equal to one or zero. This meant that the borders of the critical strip could be ruled out as a roosting place for zeta zeros. Not quite a squeeze, but the subtlest of pinches. Then, slowly, but seemingly surely, evidence began to accumulate and the squeeze was on.

Playing the percentages

In 1914, the mathematician G. H. Hardy (1877–1947) of Cambridge took a big first step. Hardy was an avid cricketer, and an almost rabid proponent of the importance, beauty, and necessity of pure mathematics. He proved that an infinity of zeros in the critical strip are indeed on the critical line. Nevertheless, it could still be that there were also an infinite number of zeta zeros off the critical line!

That same year, further progress came from a collaboration between the German number theorist (and student of Hilbert's) Edmund Landau (1877–1938) and Harald Bohr (1857–1951), brother of the Nobel Prize–winning physicist Niels Bohr. Bohr and Landau took up the study of a related question, asking what might be said about the proportion of zeta zeros in the critical strip that could be off the critical line, i.e., what was the "chance" that the Riemann hypothesis was false?

As early as 1905 a German mathematician, Hans von Mangoldt (1854–1925) had confirmed another of the unproved statements in Riemann's paper, proving that as we move up the critical strip, the zeta zeros become increasingly common, effectively waning and then waxing, in antithesis to the waxing and waning of the primes. More precisely, von Mangoldt proved that the number of zeros to be found in the critical strip up to a given height was approximately equal to the height times the logarithm of the height. Bohr and Landau asked what percentage of these could be off the critical line. In particular, they showed that of all the zeta zeros in the critical strip and up to a given height, the proportion that could reside beyond some predetermined distance from the critical line must get smaller and smaller the farther up the critical strip you go, and in fact this proportion asymptotically approaches zero. So, for example, if we proceed up the critical strip, computing for every height the percentage of zeta zeros that are beyond a distance of one-millionth from the critical line, this percentage gets smaller and smaller the farther and farther up the critical strip we travel, asymptotically approaching zero. Now, let's be clear: this is not the same as settling the Riemann hypothesis. Since the higher up the critical strip you go, the more zeta zeros you find, it is still possible that some zeros do in fact pop up off the critical line. Nevertheless, it's a great result and gives a good piece of evidence for the truth of the Riemann hypothesis.

This sort of asymptotic consideration in which the average properties (in the sense of proportions) of zeta zeros are studied as you move up the critical strip turns out to be a fruitful direction of research. Landau and Bohr were able to say something about the percentage of zeros off the critical line, but in 1921 Hardy and his longtime collaborator and colleague J. E. Littlewood (1885–1977) took the other side of this approach, and instead studied the percentage of zeros in the strip that were on the line. That is, they asked the question: as you move up the critical strip, what percentage of the zeros that you encounter at any given point are on the critical line?

Confirmation of the Riemann hypothesis is equivalent to the statement that no matter where you stop, 100 percent of the zeros are on the line. While Hardy and Littlewood were not able to deliver the whole kit and caboodle, they did make an initial breakthrough by showing that up to any height, the percentage of zeta zeros within the critical strip that are also on the critical line must be greater than zero. Furthermore, the actual percentage that they could prove depended on where you stopped to take the measurement. Unfortunately, the strength of the implication of their argument diminishes in almost direct relation to the distance up the critical strip at which it is applied. For they could prove only that as you travel up the critical strip the proportion of zeta zeros guaranteed to lie on the critical line becomes smaller and smaller, ultimately falling off to such a degree that, asymptotically, all you can conclude is that at most zero percent of the zeta zeros in the critical strip are actually on the critical line.

Nevertheless, Hardy and Littlewood's result made another substantial chink in the armor of the Riemann hypothesis. This is probably Hardy's most famous important mathematical contribution to the story of the Riemann hypothesis, but he also has another, decidedly nonmathematical, connection. Although an avowed atheist, Hardy was so fearful of travel across the English Channel that when forced to do so, he would hedge his atheistic bet by mailing postcards to friends claiming that he had a proof of the Riemann hypothesis. Hardy saw this as extra insurance for his safe passage, as he was sure that in the unlikely event that God did exist it would be a deity so concerned with mathematical progress that it would never let Hardy die leaving in doubt the veracity of his claim of a proof of the Riemann hypothesis.

Two cautionary tales

All this progress filled the mathematical world with optimism, and the details of Riemann's hypothesis seemed to be in sight, but then mathematicians received reminders, on both the theoretical side and the computational side of the coin, that until you have a proof, anything can happen.

The Limits of Computation

Each explicit calculation of a zeta zero in the critical strip confirmed the Riemann hypothesis, but Hardy had demonstrated that there were an infinite number of these zeros, so that each such confirmation is like a drop of water in an infinite well. Such are the limitations of this sort of computational approach to the Riemann hypothesis. Sure, lightning might strike and one of these calculations could turn up a stray zeta zero living off the critical line and thereby disproving the Riemann hypothesis. But no matter how many zeros we find on the line, there are still an infinite number left to check, and the danger of a counterexample looms as a possibility for each and every one of this infinity of remaining instances.

Indeed, it was in 1914, coincidentally with the work of Bohr and Landau, when the computer was but a gleam in the collective scientific eye, and just before zero-checking would begin in earnest, that mathematicians were given a less than gentle warning of the pitfalls that might lie ahead for those who were arrogant or lazy enough to rely on computational checks when the behavior of the primes is in question.

At this time Littlewood had undertaken a study of Gauss's estimate of the prime counting function. Gauss had been the first in a long line of mathematicians to make detailed comparisons of his estimate with the exact count. In all these calculations (proceeding well into the millions) Gauss's estimate for the number of primes always undercounted the exact number. So it was natural to assume that this was always the case. But in 1914, Littlewood stunned the mathematical world by showing that in fact the two counts—Gauss's estimate and the exact count—would forever run neck and neck, trading the lead infinitely often. Sometimes Gauss would overestimate, and sometimes he would underestimate, so that the graphs of these two counts entwine each other like two infinitely long and overly friendly serpents.

Littlewood proved this fact without actually indicating a place where the first of these crossings would occur. This he left to a student of his, S. Skewes of Cambridge University. In a remarkable and difficult analysis Skewes was able to determine a marker before which it was guaranteed that Gauss's estimate would overtake the exact count of the primes. In honor of Skewes's discovery, the first place where this does occur is called the ***Skewes number***. In 1933 Skewes showed that the Skewes number is at most $10^{10^{10^{34}}}$.

This is an astoundingly large number. In order to provide a few benchmarks, let's keep in mind some other rather big numbers. The current best estimate of the age of the universe is around 15 billion years, that is, about 10^{17} seconds, a number represented by a one followed by seventeen zeros. This pales in comparison with the number of protons in the universe, estimated by our friend Hardy at 10^{80}, or a one followed by eighty zeros.

To unwrap the magnitude of the Skewes number from its heliotropic notation, we work our way down the exponents, from top to bottom. Let's first consider a manageable but analogous example. When we write the number 10^{3^2} this means first take three and raise it to the second power, giving nine. Then we are meant to take ten, and raise it to the ninth power, giving us the number 1 billion, or about one-tenth of Bill Gates's fortune. Turning our attention to the Skewes number, we start with 10^{34}, the product of thirty-four tens, which is a one followed by thirty-four zeros. This already puts us in the stratosphere, somewhere between the number of seconds the universe has been around and the number of protons it contains. So far, so good. Continuing, the number $10^{10^{34}}$ is then that number represented by a one followed by 10^{34} zeros. No, don't even try to think about this. It is a number so large that even if someone had been writing one zero every second, or for that matter, ten zeros each second, since the Big Bang beginning of the universe, the transcription of this number would still not be complete. And that is just the writing—the actual number is larger still. We take this ridiculous number, $10^{10^{34}}$, and use it as the last exponent, raising ten to it, so that it represents the number of zeros that follow one in the Skewes number. We're far from the realm of comprehension or analogy. Such is mathematics, able to go where no man will ever go, in either time or space.

Hardy remarked that the Skewes number is "the largest number which has ever served any definite purpose in mathematics." Beyond comparing it with the number of protons in the universe, he also noted that it swamped the number of possible chess games, which he estimated to be on the order of $10^{10^{50}}$, a one followed by only 10^{50} zeros.

In fact, the Skewes number we've just discussed is also called the first Skewes number, for there is indeed a second Skewes number. The former corresponds to the comparison of Gauss's estimate and the acutal count, should the Riemann hypothesis be true, and the latter to the case in which the Riemann hypothesis is (heaven forbid!) false. Skewes's initial bound of the second Skewes number is even more astronomical than the first, weighing in at a whopping $10^{10^{10^{10^{3}}}}$.

Since Skewes's announcement, mathematicians have been slowly whittling down their estimate of the first Skewes number, and while there is some debate, it is known to be less than a measly 10^{1167} (according to the mathematicians R. Guy and J. Conway), so that it is at last known to be less than the number of possible chess games, but still possibly greater than the number of protons in the universe.

True, False, Neither?

Results like those obtained by Littlewood and Skewes served as a reminder to the mathematical community of the possibility that even given all this initial evidence, the Riemann hypothesis could still be false. True or false, those would appear to be the options. This is what was behind Hilbert's clarion call; he had posed these twenty-three problems, and now it was the job of mathematicians to dig into these subjects, separating truth from falsehood, like gold from dross. There is no *ignoramibus*—problems have solutions, and mathematicians will find them. The Riemann hypothesis would not be an exception.

This seemingly either-or nature of mathematics is what initially attracts many people to the subject. As children we are given pages of arithmetic problems. We work away industriously, secure in the knowledge that there is a right answer, pleased when we achieve it, and ready to go back to the drawing board when we are wrong. Mathematics provides a comfortable island of black and white in a world which is a muddle of shades of gray. When Biff Loman fails an exam in Arthur Miller's *Death of a Salesman,* it's a math exam. It's the only exam where Willy Loman couldn't invent an excuse for failure. It's the only subject capable of standing up to his self-delusion.

So strongly did Hilbert believe in the either-or ethos that he made it the basis of the second of his twenty-three problems. Hilbert wanted nothing less than to remake the foundations of mathematics. So he begins his statement of problem two with a challenge to the mathematical world to "set up a system of axioms which contains an exact and complete description of the relations subsisting between the elementary ideas of that science." Indeed, Hilbert thought that he had such a collection of self-evident statements in hand, and so he asked mathematicians to take up the task of proving "that they are not contradictory, that is, that a definite number of logical steps based upon them can never lead to contradictory results." With these statements Hilbert put forth the problem of proving the *consistency* of a set of axioms for mathematics.

The assured stability that Hilbert sought to find in the mathematical world was at odds with the way things were going in the world at large in the 1930s. Not certainty but uncertainty seemed to be the watchword of the day. A young German physicist, Werner Heisenberg (1901–1976), would show that sure knowledge of velocity and location was impossible on the microscopic scale, an "uncertainty principle" that seemed to be mirrored macroscopically in the swirling tempest of disquiet slowly spreading through Europe. As the world spun out of control, even the usually orderly world of mathematics could not provide refuge from the coming storm. For as one young Austro-Hungarian demagogue began a revolution that would cast into doubt the very foundations of human nature, simultaneously another Austro-Hungarian, the mathematical logician Kurt Gödel (1906–1978), was shaking the foundations of mathematics.

In 1931, in a landmark paper, "On Formally Undecidable Propositions of *Principia Mathematica* and Related Systems," Gödel set mathematics on its ear as perhaps no one before Cantor had done. Gödel dashed Hilbert's hope for certainty by proving that in any axiomatic mathematical system with enough "power" to do arithmetic—that is, with enough expressive ability to voice the laws of basic arithmetic and number theory—the consistency of the system can *never* be proved within the system in a finite number of steps. But even more shocking were its assertions regarding the power embedded within the axioms as well as the assumption of finite proofs: that as long as the axioms are consistent, then there will always be statements which can be neither proved nor disproved, that is, such a consistent system is not *complete,* and *undecidable* statements exist. Could it be that the Riemann hypothesis is one of these undecidable propositions?

Gödel was eventually driven from his homeland. With the help of Einstein and others, he was brought to Princeton to take a position at the Institute for Advanced Study, where after several stints as a visitor and member, he was eventually appointed to the permanent faculty. Gödel continued his mathematical and philosophical investigations, and even did important work in theoretical physics in which he discovered the mathematical possibility of universes beyond our own that are still consistent with Einstein's equations for general relativity. Although able to uncover and manage the inconsistencies embedded in science, he increasingly had trouble navigating those of life. Always a hypochondriac, as he grew older Gödel also became overwhelmed by paranoia, until he found himself inextricably tangled in a logical paradox whereby he starved himself to death from fear of being poisoned.

A GOOD CHANCE THAT THE RIEMANN HYPOTHESIS IS TRUE

In his foundational work Gödel raised the possibility that the Riemann hypothesis might never be proved, thereby bringing a sense of uncertainty to the degree to which we might ever know the primes. Meanwhile, another mathematician, or more precisely a mathematical statistician, Carl Harald Cramér (1893–1985), was instead seeking to exploit the mathematical tools that make possible the quantification of uncertainty in order to advance the study of the primes.

Cramér, who was Swedish, came to mathematics by way of chemistry. As a chemistry student in the early twentieth century, he was surely well aware of the probabilistic and statistical outlook that was then reshaping his science. Tools from the mathematical fields of probability and statistics were helping to make precise the quantum physics of the inner space of the atom. Related mathematical techniques are also at the heart of statistical mechanics, a discipline which seeks to describe how larger-scale properties of matter emerge from a combination of many small-scale atomic worlds. The power of these ideas seems to have made a deep impression on Cramér, for he eventually abandoned his chemistry studies for mathematics. Ultimately he completed a doctoral dissertation on a particular class of the Dirichlet L-series, the zeta function–like series that help us to understand periodic structures in the primes.

Cramér is often credited with having founded the discipline of *probabilistic number theory,* which uses the tools of uncertainty to make rigorous statements concerning the behavior of what would appear to be the anything but uncertain integers. This subject takes on questions of the form "What can we say is true on average about the primes or the integers?"

So Cramér looked back at Gauss's early conjecture on the growth of the primes and gave it a probabilistic spin. Gauss's computations had led him to hypothesize that, on average, the number of primes close to a given value was approximately the reciprocal of the logarithm of that value. For example, Gauss would guess that for numbers that are near 1 million roughly one over the (natural) logarithm of 1 million, or about seven-hundredths of them, are primes. In other words, around seven out of every 100 numbers near 1 million are prime. Colloquially, we might say that in blindly choosing an integer near a fixed position on the number line ("choosing at random"), our chance of coming up with a prime would be approximately one out of the logarithm of the original position.

Cramér used these findings to create a model of the primes as the cosmic outcome of a simple chance process, as if Kronecker's natural numbers-giving God had labeled each as prime or composite according to the result of the toss of a special number theorist's coin: this cosmic coin-tosser would use a different prime-composite coin for each natural number, weighted so as to come up "prime" a fraction of time equal to the reciprocal of that number's logarithm, and "composite" the remaining fraction of the time.

In this way the actual labeling of the natural numbers as prime or composite can be viewed as the result of a single run of the labeling game designed by Cramér. To distinguish the results of another play of the game from the real deal, we'll declare the outcome of a run of this game to be a labeling of numbers as either *Cramér primes* or *Cramér composites.* It goes like this: two is always a Cramér prime. Starting with three, things get a little chancy. The "three-coin" is weighted in such a way that approximately 91 percent of the time it comes up "prime" (since the reciprocal of the logarithm of three is about ninety-one hundredths) and the remaining 9 percent of the time it comes up composite. So, we flip our three-coin, and label three with the outcome. Next we move over to four, and take out our four-coin. The four-coin is weighted in such a way that approximately 72 percent of the time it comes up "prime" and 28 percent of the time it comes up "composite." We flip the four-coin and label four with the outcome. So in this model, it is possible (even likely) that four will be labeled a prime. We continue in this way, integer after integer, constructing a sequence of Cramér primes. Our familiar primes are but a single possible outcome of the game.

What sorts of properties can we expect of sequences of Cramér primes generated in this flippant fashion? Sure, there will always be the outlandish exception—like a run of luck that labels every number a prime, or its mirror image that labels each a composite. But we want to know what the usual behavior is. What sorts of properties can we count on?

As a first step, Cramér showed that almost every play of this infinite game generates a sequence of Cramér primes satisfying the Prime Number Theorem. This means that, with only a relatively insignificant collection of exceptions, for a given value (say 1 million), a run of coin tosses in Cramér's game assigns the label *Cramér prime* to a number of integers roughly equal to that value divided by its logarithm. For example, we could expect on most runs of the game that approximately 72,000 integers less than 1 million are labeled Cramér primes. So, as far as the Prime Number Theorem is concerned, Cramér primes behave just like our usual prime numbers. But even more significantly, Cramér showed that for Cramér primes, the error-term analogue of the Riemann hypothesis is true. In other words, for almost every sequence generated in this chance way, the difference between the precise number of Cramér primes and the asymptotic count of the

Prime Number Theorem is small enough that an analogous Riemann hypothesis stated for Cramér primes would be true.

In short, Cramér showed that for just about any sequence of labelings that bear some statistical similarity to the intermingled sequence of primes and composites that are our own natural numbers, the corresponding Riemann hypothesis would be true. The rub is that we don't know a priori for which plays of the game the Riemann hypothesis is false and for which is it true. We know only the result in aggregate; this is like knowing that almost any infinite game of tossing a fair coin will have equal proportions of heads and tails, but we can't know before we start any particular game if it will be one of those remarkable ones in which the tails greatly outweigh the heads or vice versa.

Cramér went on to prove that in almost all the "Cramér number worlds," those properties for our ordinary primes known to be consistent with the truth of the Riemann hypothesis were also true when rewritten as properties of Cramér primes. This shows that at least in the probabilistic worlds cooked up by Cramér, things looked just as mathematicians were expecting them to be in the world of our usual primes, assuming that the Riemann hypothesis turned out to be true. So, the Riemann hypothesis was looking like a pretty good bet.

Evidence accumulates

In 1932 Carl Ludwig Siegel (see Chapter 6), formerly a student of Landau's, discovered a mathematical language better suited for some aspects of studying the Riemann hypothesis. Until this time Riemann's zeta function had been studied mainly through its so-called integral form. This had been useful for various aspects of theoretical analysis and in particular enabled de la Vallée-Poussin's proof of the Prime Number Theorem. But with respect to computational questions in which you try to see what the sum of reciprocals raised to a given complex number actually produces, the integral form was opaque. In mathematics, as in poetry, some choices of language are better suited than others for certain purposes. Different formulas for a given quantity or function are like different metaphors, capable of revealing different aspects of a single truth. The *Riemann-Siegel formula* would be a better way to say "The zeta function produces this."

Siegel uncovered the basic idea behind the Riemann-Siegel formula while working through Riemann's unpublished working diary, or *Nachlass.* Approximately 100 pages of this *Nachlass* contain notes or calculations related to the Riemann hypothesis. In his book *Riemann's Zeta Function,* the de facto bible for working mathematicians interested in the Riemann hypothesis, H. Edwards remarks that at least several other first-rate mathematicians had tried to make sense of the *Nachlass* and failed, but Siegel persevered.

Siegel states that the *Nachlass* contains no steps toward a proof of the Riemann hypothesis. Nevertheless, in a published account of what he found in the *Nachlass,* Siegel left no doubt that the formulation of the Riemann hypothesis was not a lucky guess on Riemann's part but rather the result of extremely careful work and detailed calculation, along with great insight. In an astounding achievement of forensic mathematics, Siegel succeeded in performing a thorough analysis of Riemann's various jottings and calculations, and from this managed to derive what he believed was the formula that Riemann used for computing the zeros of the zeta function. The result was the Riemann-Siegel formula.

The discovery of the Riemann-Siegel formula made possible a raft of new tricks for studying the Riemann hypothesis, and in particular greatly expedited the calculation of zeta zeros. By the late 1930s, the mathematicians E. C. Titchmarsh (1899–1963) and L. J. Comrie (1893–1950) had completed a calculation of the first thousand zeros and had found them all to be on the critical line.

World War II and a Riemann Hypothesis That Is True

World War II interrupted or cut short many a life, academic or other. Like many of their brethren, European scientists and mathematicians of Jewish descent, if lucky enough to escape the Nazis, fled their homes and jobs.

Both Siegel (who was not Jewish, but openly despised the Nazis) and Hadamard left for the United States. Siegel took up a position at the Institute for Advanced Study in Princeton, New Jersey. In spite of Princeton's peaceful and intellectually stimulating atmosphere, Siegel "never really felt comfortable" in the United States, calling his stay a "self-imposed exile," and he looked forward to the day when he could return to his home in Berlin. Hadamard, at that time seventy-five years old, could not find a permanent position, but he was able to come to New York with his family and assume a visiting professorship at Columbia University.

Almost all scientific minds around the globe were either volunteering for, or being pressed into, work directly relevant to military purposes. Nevertheless, either in spite of the war or perhaps as a respite from it, some pure science continued, including work on the Riemann hypothesis.

In 1942, the Norwegian Atle Selberg (b. 1917) pushed ahead in the direction of work initiated by Hardy and Littlewood. Selberg showed that asymptotically, the proportion of zeros known to be in the strip and on the critical line is positive. Hardy and Littlewood had fallen just below the asymptotic postivity barrier: at any given height of the critical strip they could produce a positive fraction of zeros, but as the height grew their estimate continued to get even closer to zero, or a result that is asymptotically zero.

For his achievement, as well as for a new proof of the Prime Number Theorem, accomplished with the help of the peripatetic Paul Erdös (1913–1996), Selberg, along with French mathematician Laurent Schwartz, would receive the Fields Medal in 1950. The Fields Medal, made of pure gold, is named in honor of the Canadian mathematician J. C. Fields, who endowed the prize in 1932. It is the most prestigious prize given in mathematics, and thus is often referred to as the "Nobel Prize of mathematics." One way in which it differs from the actual Nobel Prize is that the latter is generally bestowed late in a scientist's career, marking a fundamental discovery that has stood the test of time, whereas the Fields Medal is by convention given to as many as four of the best and brightest mathematicians younger than forty—after all, it is in the nature of mathematics that their theorems are guaranteed to stand the test of time.

Various stories claim to explain why no Nobel Prize is awarded in mathematics. One of the best is that Alfred Nobel was cuckolded by the Swedish mathematician Gösta Mittag-Leffler, and thus excluded mathematics as a bittersweet revenge. The much less romantic, but (by most accounts) apparently true reason is that pure mathematics was simply not one of Nobel's scientific interests.

Ever since 1936 the Fields Medals have been awarded at the International Congress of Mathematicians (ICM). The first ICM was held in 1893, and it has evolved (like the Olympics) into an event held every four years.

Like many international events, the ICM was interrupted during World War II, so that the meeting of 1950 was the first since 1936. It was still scarred by geopolitical turmoil, as the Soviet Union's mathematicians were forced to stay at home. The head of the prize committee in 1950 was the Danish mathematician Harald Bohr, who together with Landau had at the turn of the century helped refine our understanding of Riemann's zeta function. Bohr remarked that Selberg's work on the zeta zeros was evidence of "his extraordinary penetrating power." Like Einstein, Gödel, and so many other great thinkers, Selberg would also eventually obtain a permanent position at Princeton's Institute for Advanced Study.

A result like Selberg's continued the process of applying the pincers of logic to the critical strip, but it is fitting that perhaps the strongest evidence of hope for trapping the zeta zeros on the critical line was discovered in the unlikely location of an internment camp in France. There, awaiting trial for desertion, was a young French mathematician, André Weil. In the quiet of the camp, Weil had been working on settling a different Riemann hypothesis, one inspired by the original, and this one had turned out to be true.

A "ZOO" OF ZETA FUNCTIONS

Riemann had been able to reveal the importance of Euler's proof of the infinitude of primes by seeing the harmonic series as a single example of an entire complex plane's worth of infinite series. It might thus follow that in order to settle the Riemann hypothesis, perhaps we should try this trick once more, a meta-trick if you like: find a way to view Riemann's zeta function as but one of an infinitude of zeta functions, each with its own "Riemann hypothesis." Perhaps a distantly related cousin might reveal a previously hidden aspect of the original Riemann zeta function, and thus shed some light on a path toward the resolution of the original Riemann hypothesis. In this manner we see abstraction as both light and shadow, occluding differences through the scrim of intellect in order to illuminate some general principle whereby the particulars unite into a whole.

Eventually Riemann's zeta function would be seen not just as a singular oddity used to understand the distribution of the primes, but rather as one species in a number-theoretic jungle, growing into what is today a veritable "zoo" (in the words of the mathematician Peter Sarnak of Princeton University) of zeta functions, bearing varying degrees of resemblance to Riemann's zeta function, which for mathematicians everywhere remains *the* zeta function.

The first generation of these new zeta functions, each with its own "Riemann hypothesis," seems not so far removed from its original ancestor. The complexified L-series of Dirichlet, and even the ideal number-inspired zeta functions of Dedekind, clearly share a family resemblance. Each possesses an associated critical strip whose concomitant Riemann hypothesis still conjectures that all the new nontrivial zeta zeros lie on a critical line, indeed that same critical line of the original, now assumed to be crowded with zeros (thanks to Hardy, Littlewood, Selberg, and others).

However, the resolution of any of these more general Riemann hypotheses appears just as tough as the original. Perhaps in our search for insightful generalization we have stopped too soon. Remember that Euler did not go far enough when he considered only series of reciprocals raised to integer powers. Nor did Dirichlet go far enough when he permitted real exponents. We needed Riemann's intellectual bravado to see the utility of working with an exponent that was a complex number in order to get to the heart of the distribution of the primes. We needed Riemann to see that the way to the Prime Number Theorem would go through the complex plane. What of the Riemann hypothesis, then? Is there a new space in which we might travel, and a new Beatrice to guide us through it, in order that we might find a Riemann hypothesis that is provably true?

There is. It is a new, wild kingdom of number and function called *function fields,* and in this unfamiliar place resides a zeta function whose Riemann hypothesis would be proved true. Weil was the intrepid explorer who remade analysis here.

ANDRÉ WEIL

André Weil (1906–1998) was born into a prosperous, intellectual Jewish family in France. He was the brother of Simone Weil, the famous though tragic social activist and moral philosopher. A precocious child, Weil soon showed a tremendous talent for languages and mathematics. A scholar's scholar, he was well known for his historical outlook on mathematics, and he would often stun (and sometimes irritate) his colleagues with his ability to trace almost any piece of work to its true inspiration and origin.

Weil is considered by many the father (or perhaps by now the grandfather) of "modern" algebraic geometry, the twentieth-century remaking of the first introduction of algebra into geometry through coordinates. By giving geometric life to equations, algebraic geometry enables the application of geometric intuition and techniques to the problems of number theory. Among these are the *Diophantine problems* (named after the Greek mathematician of the third century, Diophantus) that are part and parcel of early algebra courses. A typical Diophantine problem is "find two integers whose sum is 20, and whose squares sum to 208." The former condition can be interpreted as requiring that the two numbers mark coordinates on a line of slope of minus one that crosses the vertical axis at a position twenty units above the origin; the latter condition demands that the coordinates of the solution also lie on a circle whose radius is the square root of 208, centered at the origin. Today, elaborate families of equations with thousands of unknowns arise in the ubiquitous problems of resource allocation. It would be impossible to solve these problems if not for related geometric methods.

Weil not only remade algebraic geometry but was even so bold as to attempt a Hilbert-style reinvention of all of modern mathematics through his creation of a group called the Séminaire Bourbaki. This was a secret mathematics society which periodically issued encyclopedic texts (written by a committee) on the most important areas of advanced mathematics. Each book or report was published under the pseudonym Nicholas Bourbaki, a mythical creation of Weil and his wife, Eveline, named for a Napoleonic general, Charles Bourbaki, who had suffered one of France's most memorable military defeats. The pretense was sustained even to the point of creating a biography for Nicholas, which included his birth and citizenship in the fictional country of Poldavia. Weil was so enamored of the fiction that in his own curriculum vitae he mentioned membership in the Poldavian Academy of Sciences. The Séminaire Bourbaki lives on today, less secret, but still of great stature, issuing reports and papers that provide an important source of mathematical research, written by many of the world's leading mathematicians.

When World War II broke out, Weil was touring Finland. He decided to try to wait out the war there but was arrested after the Russians invaded. If not for the eleventh-hour intervention of the mathematician R. Nevanlinna (1895–1980), Weil would have been executed as a spy. Instead, he was deported back to France, where he was imprisoned in Rouen as a draft dodger. There, under the constraints of prison life (although he was able to receive books, papers, and visits from friends and colleagues), he devoted his intellect to, and eventually succeeded in, settling the *Riemann hypothesis for function fields.*

The Riemann hypothesis for function fields is a result smack-dab in the middle of modern algebraic geometry, concerning itself with the solutions of basic equations but ones in which the "numbers" that you are allowed to plug in are not the usual integers or even complex numbers. Instead, these equations are meant to be solved using elements of *function fields.*

If the natural numbers and their expansions into the integer, rational, real, and even complex numbers are mathematics in vivo, serving as they do as the language of physics and descriptors of the world, then function fields are almost a mathematics in vitro. Every prime number has an associated function field that is in essence an analogue of the rational numbers. Integers are replaced by basic geometric objects called curves, but not quite the curves that are the graphs of high school geometry. Those familiar curves can be traced out by continuous pencil lines in our looseleaf notebooks, effectively creating an association between one infinite set of real numbers and another set that are the points on the curve. But the curves of function fields work in a different setting, that of *finite fields,* in which we can use our early geometry only as intuition.

The simplest example of a finite field comes from a number system that some of us learn as "clock arithmetic" but that might better be called "odometer arithmetic." On an odometer with only one wheel, showing the digits zero through nine, if you start at zero, then travel four miles, and then another nine miles, the odometer will read 3. More generally, in (one-wheel) odometer arithmetic, the number displayed is the remainder obtained when dividing the total distance traveled by ten. Going one distance and then going a bit more gives addition in odometer arithmetic, and similarly you can perform multiplication: take two numbers between zero and nine, multiply them, and then divide the product by ten and take the remainder. So far, so good. The problems start when you try to do division. In order for this to make sense, you need an odometer that doesn't repeat itself after ten clicks but instead has a prime number of settings.* As long as this is true, then you have a number system that allows for addition, subtraction, multiplication, and division, just like the rationals. But it is finite. This is a finite field.

Whereas Riemann's zeta function provided a way to count the primes less than a given magnitude, the zeta functions relevant for finite fields instead count the number of points with particular properties that are on these curves defined over finite fields. Associated with this zeta function is a Riemann hypothesis, and it was this Riemann hypothesis that Weil was able to wrestle to the ground.

Although Weil did most of the basic work on this problem while he was in prison, it was not written up in detail until some time later, after he was released and then emigrated (in 1941) to the United States. In the preface to his collected works, Weil writes that his future colleague Hermann Weyl once offered to have him put in prison again if it would help him complete his work on this topic.

In finding a proof for the Riemann hypothesis for function fields, Weil added another piece of evidence to support the possibility of bringing a similar resolution to Riemann's original conjecture. Soon

*Notice that in our decimal odometer arithmetic, five times two gives ten, which then gives zero (after taking the remainder upon division by ten). So it's possible for two nonzero numbers to multiply together to produce zero. If instead of using ten, we do our arithmetic on an odometer with a prime number of settings, this cannot happen. This is the key to being able to perform division.

after his great achievement, Weil, like Selberg and Siegel, settled into a position at Princeton's Institute for Advanced Study.

Close, but no cigar

The Institute for Advanced Study slowly became the place to be for anyone studying zeta zeros. Accordingly, in the spring of 1945 the German mathematical analyst and number theorist Hans Rademacher (1892–1969), then a professor at the University of Pennsylvania, arrived in Princeton, holding what he believed to be a disproof of the Riemann hypothesis. In spite of the growing collection of evidence to the contrary, Rademacher thought he had an argument showing that eventually one of the nontrivial zeta zeros would be found off the critical line. With the draft of a proof in hand, he had traveled to Princeton to see his old friend Siegel, the reigning expert on the Riemann hypothesis. Together they pored over the manuscript, and by day's end they had agreed that here was a disproof of the hypothesis. Satisfied (and almost surely, giddy with excitement), Rademacher returned to Philadelphia.

According to Selberg, after Rademacher left, Siegel began to feel somewhat uneasy about the discussion; later that evening, Siegel returned to the manuscript to check it once again. Finally he found what had been bothering him: Rademacher had made a classic mistake, using a *multivalued function* as though it were single-valued. In a different version of the story, Rademacher called Siegel several days later to retract the paper after waking up in the middle of the night shouting, "It is wrong!"

"You need to be very careful when working with multivalued functions," warns Selberg as he recalls the event. Indeed, the analysis of multivalued functions is delicate business, the subject of another of Riemann's mathematical contributions, the invention of *Riemann surfaces.* Everyday *single-valued functions* are precisely like the zeta function PDAs discussed earlier, machines that take a number in, transform it according to some rule, and then spit out the new transformed number. One goes in, and one comes out. A multivalued function is such that multiple outputs are possible; therefore, when using such a souped-up mathematical machine, you also need to designate beforehand which of the outputs is of interest. The pictures associated with multivalued functions can be complicated but beautiful corkscrewlike objects. The outputs are grouped into "sheets" interwoven like spliced stacks of pancakes, each pancake corresponding to a different collection of function values, or "branches." The design flaw that led to the famous collapse of the Tacoma Narrows bridge is said to have resulted (in part) from using the wrong branch of the most elementary of multivalued functions, the logarithm. Rademacher's work on the Riemann hypothesis collapsed because of a similar oversight.

Ordinarily, this would not be a big deal—a personal disappointment for sure, but that would be the end of it. In the course of research, mathematicians make mistakes all the time, but these missteps usually never leave the study. On occasion they make it to the hands of a journal editor, who with the help of anonymous experts or referees takes on the job of checking the paper for errors as well as gauging the level of significance of the work (is it "journal-worthy?"), an evaluation which will vary according to the status of the journal. There is a long history of this tried, and usually true, process of peer review, by which mathematics, and science in general, keeps moving forward, ensuring a foundation of truth every step of the way.

Sometimes errors are found, and papers are returned. Authors attempt to fill gaps in reasoning and emend erroneous conclusions, and then resubmit their corrected work. Usually, no one hears about the refereeing process; but for problems of great magnitude, things can be different. For example, when, several years ago, Andrew Wiles announced a proof of Fermat's Last Theorem, it was front-page news in the *New York Times.* A special committee was convened to study the manuscript, and ultimately discovered an error, which—after a year or so of further work in collaboration with Richard Taylor (then of Cambridge University, now of Harvard)—Wiles was able to fix.

Rademacher's story has a less happy ending. The chairman of the mathematics department at the University of Pennsylvania was something of a publicity hound, and without Rademacher's permission he had announced the result to journalists. In fact, the chairman had gone so far as to ask the editors of *Transactions of the American Mathematical Society* to hold the presses while the paper was refereed. Articles ran in *Time* magazine as well as the *New York Times.* Unfortunately, there was no rescuing Rademacher's result. Rademacher's proof (or disproof) was dead, and the Riemann hypothesis was still alive and kicking.

Rademacher was terribly embarrassed by the ordeal and never spoke of it again. He had a summer home near Dartmouth College, at whose mathematics department he was a frequent visitor. It was well known in the department that no one was to mention the words "Riemann hypothesis" in his presence.

Pushing forward

The successes of Weil and Selberg, and even Rademacher's misstep, reflect an evolving understanding of the Riemann hypothesis. Theoretical progress was matched by a growing body of evidence made possible by the advent of the modern computer. Computing machines had been developed during World War II, when they were necessary for the quick execution of the many calculations required by such urgent wartime tasks as computing missile trajectories, designing airplane wings, and ultimately unraveling the mysteries of nuclear physics relevant to the design of the atomic bomb. Day by day, the most important military role played by the computer was as a code breaker. Among the greatest of the human code breakers (cryptanalysts) was Alan Turing (1912–1954), who would try to turn his skills in code breaking and computer building to cracking the code of the Riemann hypothesis.

Turing is arguably the father of the discipline known today as *theoretical computer science.* This is the investigation of the abilities and limits of computation, often embodied in the analysis of a problem's *computational complexity,* a measure of the number of steps a computer will need to arrive at its solution. Turing was the creator of the *Turing machine,* a "universal computer" which still serves as the abstraction through which the theoretical efficiency of any computer program is analyzed. Turing's careful analysis of the power inherent in this machine is the origin of *complexity theory.* Of equal renown is his creation of the *Turing test,* a procedure proposed as a benchmark for the advent of intelligence in a computer. In this game an interrogator types at a keyboard, engaged in dialogue with an unknown respondent. His goal is to determine through this electronic conversation if he is communicating with a machine or a person. Turing and others believed that should a machine "pass" the Turing test by being identified as a person, it should be viewed as having "intelligence." Turing's primacy in the world of computer science is recognized in the annual presentation of the Turing Award by the Association for Computing Machinery (along with a cash prize of $100,000). The Turing Award is considered the Nobel Prize of computing.

Turing is perhaps most famous for cracking the German Enigma codes during World War II. But just before Turing turned to wartime work, he had been wrestling with the problem of cracking the code of the primes. He tried to find a more realistic estimate of the Skewes number, and he was also thinking about trying to solve the Riemann hypothesis.

As witnessed by Turing's success in breaking the Engima codes, he was not simply a blue-sky academic. Some computer scientists might tell you that *in theory* the Enigma code could be broken—but clearly that would not do. Similarly, no Turing machine would do in order to study the Riemann hypothesis. Rather, for this problem Turing had in mind building an *analog computer* dedicated to the computation of zeros of the zeta function. This was a machine based on physical principles that would mimic the physics embedded in the Fourier expansion of the zeta function. The name *analog* is derived from this act of mimicry, or analogy. The design of the zeta-function computer was inspired by an already functioning analog computer in Liverpool, which, through a similar mechanical analogy, was being used to compute the schedule of the tides. In 1939 Turing received a grant of £40 from the Royal Society to build his machine. In the grant application he wrote, "I cannot think of any applications that would be connected with the zeta-function."

Turing's special-purpose Riemann hypothesis computer was designed but never built. After the war Turing took up the problem of the zeta zeros once again, and in 1953, using a more conventional computer, he devised an algorithm based on the Riemann-Siegel formula to begin computing nontrivial zeta zeros. As a result, he was able to confirm that every single one of the first 1,041 nontrivial zeta zeros was on the critical line.

Turing was the first in a long tradition of computer scientists who would build powerful computing machines and then put them through their paces by confirming the Riemann hypothesis up to some dizzying height in the critical strip. Further fine-tuning of Turing's method, as well as the invention of increasingly efficient computers, later enabled R. Sherman Lehman's computation of the first 250,000 (1966). In 1968 John Barkely Rosser, J. M. Yohe, and Lowell Schoenfeld extended these calculations to the first 3.5 million zeros.

By 1974 there were two more landmarks in the pursuit of the Riemann hypothesis. The first was a dramatic improvement over Selberg's wartime work. Selberg had been the one to crack the positivity barrier by proving that a nonzero fraction of the zeta zeros in the critical strip were guaranteed to be on the line. But Selberg's proof was of a type known as an *existence proof*—guaranteeing the existence of this nonzero number without actually producing it. Existence proofs are very frustrating to many mathematicians, because they provide rigorous proofs of the existence of a certain kind of number, or in other cases a geometric object, without providing a means of construction. However, in 1974 Norman Levinson (1912–1975) made a huge improvement by showing that as you climbed the critical strip, at least one-third of the zeros encountered up to any point were on the critical line. So now we're up to 33 percent and climbing.

Other progress followed, along the route of randomness first initiated by Cramér in the 1930s. Cramér primes were generated by a model akin to independent coin tosses, using a differently weighted coin for each natural number to label the number prime or composite according to its outcome of heads or tails. One objection to this way of modeling the appearance of primes might come from the observation that if a coin toss causes us to call a particular number a prime, then we should probably rule out all its multiples as primes. In 1958, David Hawkins (1913–2002) suggested just such a model.

Hawkins started each run of his prime lottery by first labeling two as prime and then removing from consideration all multiples of two—the even numbers. Now it's time to start tossing coins. We move on to three and, as with Cramér, we take out our three-coin; but we use a different weighting: this time it comes up "prime" one-third of the time, and comes up "composite" two-thirds of the time. We flip the coin, and if it comes up on the "prime" side, then we label three as a prime (a "Hawkins prime"), and also remove from future consideration all multiples of three. Since we have already removed four from consideration (as a multiple of two), the next number to test is five. In general, at each step we move to the next number not already removed by previous considerations. Once again we use the reciprocal of this number as the chance that the coin we flip will come up "prime"; if it does, we now remove all its multiples from consideration.

In 1974, Werner Neudecker and David Williams showed that almost every sequence of "Hawkins primes" satisfies the associated Riemann hypothesis. Once again, the ideas of randomness had been used to suggest that the Riemann hypothesis is true.

GLIMMERS OF THE NEXT GREAT IDEA

At nearly the same time as Turing was adding computational fuel to the fire, one of the saddest and most bizarre episodes in the history of the Riemann hypothesis took place. The prominence of the Riemann hypothesis was such that the brilliant, fame-seeking future Nobel laureate John F. Nash, Jr. (b. 1928), took up the challenge. In a now famous lecture given at Columbia University, Nash put forward his ideas for what was ostensibly an approach to settling the Riemann hypothesis.

When Nash was at the height of his intellectual powers, he was a formidable figure. At the time he took the podium, he was already known for what would become his Nobel Prize–winning work in game theory, as well as for having recently solved a long outstanding problem in Riemannian geometry. Sadly, however, rumors that he was mentally ill were also making their way around the mathematical community.

Nash's lecture on the Riemann hypothesis was a legendarily odd tangle of numerology and number theory, confirming for many the rumors of illness. It thus marked the public beginning of Nash's forty-year battle with schizophrenia. It has been said by some that the collision of Nash's intense concentration and obsessive nature with the sheer difficulty of the Riemann hypothesis may have contributed to the onset of the disease which incapacitated him for much of his adult life.

Most mathematicians don't believe that Nash ever had a proof, but of course it is impossible to tell. So much of science proceeds by the serendipitous juxtaposition of ideas—a stray sentence overheard at a lecture, a passing sight that engenders a waterfall of ideas. While aiming at the Riemann hypothesis, Nash was also attempting a reformulation of quantum mechanics. When Riemann conjectured his hypothesis, he too had been in the midst of a simultaneous consideration of number theory, physics, and geometry. Who is to say that at the moment when Nash's illness proved debilitating, this was not also a time when physics and number theory might have been on the verge of coalescing in his brilliant, beautiful mind in order to produce the answer to this deepest of mathematical problems?

As it turns out, precisely just such a connection—one forged between physics and number theory, found in the chance intersection of two scientists—is to be the next breakthrough in our story. It happened at the Institute for Advanced Study, which housed so many of the heroes of this tale and was a place where, periodically, Nash would find respite from his illness. The breakthrough showed that the ideas of randomness of Cramér and Hawkins, attempts to study the primes and the Riemann hypothesis as one blade of grass in a haystack of phenomena, were right in spirit. What we will see is that we were looking in the wrong haystack.

11 *A Chance Meeting of Two Minds*

THE UPHEAVAL of the first half of the twentieth century brought much of the European brain trust to the United States, and this intellectual diaspora included some of the best minds working on the Riemann hypothesis. Although the Riemann hypothesis was born in Germany and grew up in Europe, it would reach maturity in New Jersey, at Princeton's famed Institute for Advanced Study.

In the very place where Einstein plumbed the mysteries of the universe and Gödel unsettled the foundations of mathematics, the chance encounter of a physicist, Freeman Dyson (b. 1923), and a mathematician, Hugh Montgomery (b. 1944), would set the direction of research on the Riemann hypothesis through to the present, connecting the deep structure of the basic elements of numbers with that of the basic elements of matter. Within the energy levels characteristic of wave phenomena of heavy nuclei would be found a mirror of the descriptors of the primal waves that are the zeta zeros.

"THE INSTITUTE"

Better known as "the Institute," Princeton's Institute for Advanced Study is perhaps academe's most hallowed think tank. Just a stone's throw from the pastoral campus of Princeton University, abutting deer-filled woods and a local golf course, the Institute was founded in the 1930s as an intellectual boutique, funded by department store magnates, the Bambergers.

A small set of scientific luminaries made up the original faculty.

In addition to Einstein, three of the best mathematicians of the day were also lured from Princeton University: John von Neumann (1903–1957, a mathematical physicist who among many other achievements is usually recognized as the father of the modern computer and game theory); James Alexander (1888–1971, a mathematician well known for his work in knot theory), and Oswald Veblen (1880–1960, a topologist and geometer). From this original gang of four, the faculty has grown to about thirty, spread fairly evenly over Schools of Mathematics, Natural Sciences (physics), and Social Sciences.

Created as a haven for thinking, the Institute remains for many the Shangri-la of academe: a playground for the scholarly superstars who become the Institute's permanent faculty. These positions carry no teaching duties, few administrative responsibilities, and high salaries, and so represent a pinnacle of academic advancement. The expectation is that given this freedom, the professors at the Institute will think the big thoughts that can propel social and intellectual progress. Over the years the permanent faculty has included Nobel laureates as well as recipients of almost every other intellectual honor. Among the mathematicians, there have been several winners of the Fields Medal. While there are no official criteria for being offered an Institute professorship in mathematics, it does seem clear that making significant progress on the Riemann hypothesis at least puts you in the running. In this regard we've already heard of the Riemann-related achievements of three former faculty members at the Institute: Atle Selberg, André Weil, and Carl Siegel. The party has only just begun. . . .

If the permanent faculty makes up the intellectual foundation of the Institute, the lifeblood is provided by the parade of international visitors who bring a continuous influx of new ideas. They may come for as little as an afternoon, or as long as a few years, in which case they take up temporary positions as Institute "members." Their achievements and intellectual potential have earned them the precious time needed to think, write, and—perhaps most important—interact with other members as well as with the permanent faculty.

Lectures and seminars provide the usual academic venues for sharing ideas, but often even more fruitful are the frequent opportunities for informal exchange. Many visiting members choose to live in the adjacent Bauhaus-inspired Institute housing (designed by Marcel Breuer) that lines streets named for some of the most famous faculty members. In this polyglot neighborhood just off Einstein Drive, physicists, mathematicians, and historians live cheek by jowl, dropping their children off at day care, using the recreational facilities and laundry, or tromping through the acres and acres of the Institute's nearby woods. Academic posturing is a little more difficult to sustain while chasing your six-year-old.

Lunch is another time for socializing. Many members and visitors take advantage of the Institute's cafeteria, overseen by a world-class chef (who also has command of the Institute's renowned wine cellar). The spacious, stylishly modern dining room looks out on one side at a lovely enclosed garden. As the lunch hour progresses, the tables fill up, creating a veritable intellectual archipelago. For the most part the superstructure of the Institute is replicated here: a long table of physicists at one end, historians and social scientists at another table, and smaller hybridized groups scattered about. Smack in the middle, as if replicating its odd station between science and the humanities as well as its central position in the sciences, is a long table of mathematicians.

The most prominent social venue is the daily afternoon tea. On any weekday at about three o'clock, almost all the people present at the Institute arrive at the centrally located Fuld Hall, where the tearoom affords sweeping views of the Institute's grounds. Members, the faculty, and visitors mingle, taking a welcome break from hours spent huddled over a manuscript, or a day of frustration spent staring at some knotty problem. They collect for a caffeine and sugar infusion of tea, coffee, and cookies; take a moment to relax with a newspaper or a magazine; and maybe, just maybe, have a random encounter that might unravel a day's or even a year's worth of confusion. One such caffeinated collision between Montgomery and Dyson brought about what could be the single most important breakthrough on the Riemann hypothesis since the day it was first conjectured.

MONTGOMERY AND THE PAIR CORRELATION

In 1974 Hugh Montgomery was visiting the Institute. Now he is a professor of mathematics at the University of Michigan, but then he was at the beginning of his career, a recent Ph.D. from Cambridge University. He was already a number theorist of some renown, having won several prizes for his work in analytic number theory, and in particular for his study of the zeros of Riemann's zeta function.

Montgomery's analysis of the zeros had taken the form of a classic scientific gambit. He had decided to first assume the truth of the Riemann hypothesis, and with that assumption in hand, investigate its implications for the overall structure of the zeta zeros. As a general methodology this sort of approach is a tried and true research technique. One possibility is that the truth of the Riemann hypothesis may have implications that are at odds with known facts, contradictions that force us to admit its impossibility. On the other hand, it might turn out that having assumed the truth of the Riemann hypothesis, we are led to new implications which might in turn admit more easily verified hypotheses, or perhaps point to new and surprising connections. It is an outcome of the latter sort that is responsible for much of the excitement now surrounding the Riemann hypothesis.

Montgomery asked the following question: If in fact the Riemann hypothesis is true, so that all the nontrivial zeta zeros do lie on the critical line, then what can be deduced about the manner in which they would be arranged on this line?

Surely there is some structure in the seeming randomness of the zeta zeros. Von Mangoldt's work of 1905 had confirmed Riemann's assertion that the farther up the critical strip we go, the closer together the zeta zeros occur. If we imagine the zeros dropped along the critical line like sandbags from a car speeding along the center stripe of a highway, then dropping a bag at one location will influence the likelihood of dropping a bag at more distant locations. One analogy is with the prime numbers themselves: the relatively high density of small primes implies that larger numbers are less likely to be prime, since larger numbers are more likely to be multiples of earlier primes. In short, it seems that should all the zeros fall on the critical line, their positions would be *correlated.* How to quantify this intuition?

One way of measuring the degree to which a collection of numbers are correlated is by computing their ***pair correlation.*** The pair correlation measures the distribution of differences between all the pairs of numbers, or, equivalently, the distribution of distances of a set of points all of which lie on a single line. For example, a collection of tightly clustered points on a line would have the property that the vast majority of distances between points is quite close to zero; only a very small proportion (if any) of interpair distances would be large. On the other hand, points that are widely distributed might very well have the property that a sizable proportion of the distances between two points would be relatively large.

Were the Riemann hypothesis true, then the pair correlation could be used to better understand the zeta zeros. In this case, all the nontrivial zeta zeros would lie on the critical line in the complex plane; that is, all these complex numbers would have a real part equal to one-half. Thus the distance between any two points is simply the difference between their imaginary parts. After "rescaling the data"—a technical modification necessary to ensure that in comparing the pair correlation of the zeta zeros with that of other processes we are comparing apples with apples—the pair correlation is computed using the differences of the imaginary parts of all the nontrivial zeros.*

To test the assumption of the randomness (or nonrandomness) of the zeta zeros, it is necessary to have at hand for comparison's sake a paradigm for randomness. A classic model is the *Poisson process,* named for the mathematician and statistician Siméon-Denis Poisson (1781–1840). Were a Poisson process directing the dropping of sandbags on the critical line, then the chance that a sack was dropped over any given small instant of time would be proportional to the time just traveled; furthermore, the chance that two or more sacks were dropped over a short stretch would be almost zero. Most important is its defining property: that whatever happens in one small interval of time has no bearing on any later interval, often summarized by calling it a "memoryless" process. These simple rules turn out to make a very good model for a spectrum of examples of real-world randomness, including taxicab arrival times, lengths of movie queues, accident rates, radioactive decay, and meteor strikes.

Were the zeros to occur in a manner that resembled anything like

*A typical example of rescaling would occur in comparing the distances between objects in two different photographs. You could compare them by measuring directly on the photographs only if the two photos have the same scale—e.g., one inch per sixty feet. If not, then you need to "rescale" to put them on the same footing.

the perorations of a Poisson process, then their rescaled relatives would exhibit a pair correlation that is approximately constant—in particular, always near one. This reflects the fact that were the zeros simply randomly sprinkled along the critical line, the distance between any two zeros would be as likely to be one number as any

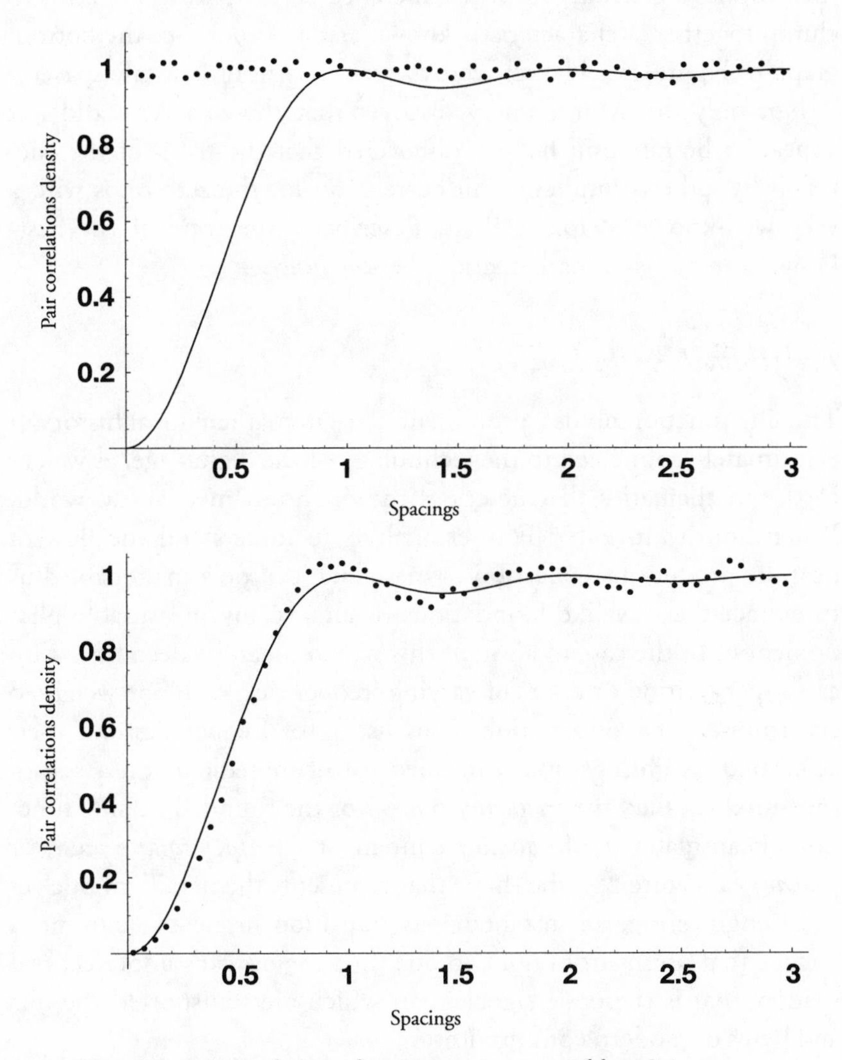

Figure 25. Pair correlations of 10,000 points generated by a Poisson process and of the first 100,000 zeta zeros (all on the critical line). In each case we compare the plots with a smooth curve built from the sinc function. The former set (top) are scattered evenly about as predicted, while the latter (bottom) hugs the curve nicely. So, the zeta zeros are not random.

other. This can be seen in the top graph of Figure 25. However, working under the assumption that the critical line was the home of all the nontrivial zeros, Montgomery found that their pair correlation would look quite different. Rather than the flat distribution characteristic of randomness, the pair correlation of the zeta zeros rises to a wavy line that implies a generally distinct reluctance on the part of the zeros to clump together, a characteristic known as repulsion. (See the bottom graph of Figure 25.) The zeta zeros were surely not randomly spaced.

Not only did Montgomery discover that the zeta zeros did not appear to be random; he also discovered that the truth of the Riemann hypothesis implies a pair correlation for the zeta zeros with a very well-known shape, a shape described using one of the best-known functions in mathematics, the *sinc function.*

Everything Is in the Sinc

The sinc function holds a prominent place in mathematical history. It is intimately connected to the technology of the digital age, as well as to the mathematics that describes the quantum mechanical world. When Fourier invented Fourier analysis to understand the flow of heat, he produced a collection of mathematical tools that enabled us to extract the wavelike foundations of almost any measurable phenomenon. In the case of a sound this would mean its decomposition as a superposition of waves of varying frequencies, each one weighted according to its contribution. This listing of frequencies and their associated weightings, like a detailed list of ingredients on a bag of cake mix, is called the *frequency content* of the sound. The sinc function encapsulates the frequency content of a perfect *square pulse,* or *boxcar* (see Figure 26), the shape that represents the paradigm of electrical engineering, an instantaneous transition in electric current, a voltage that jumps from none to one. It is the boxcar's imperfect real shadow that is the basic signal upon which are transported the bits and bytes of modern communication.

The boxcar represents the perfectly *time-limited* phenomenon: zero everywhere except for a small finite nonzero instant when it measures exactly one. In this way the boxcar represents the perfect disturbance of either sound or light, present for a moment and leaving nary a wake in its disappearance. But within its precisely delineated finitude is

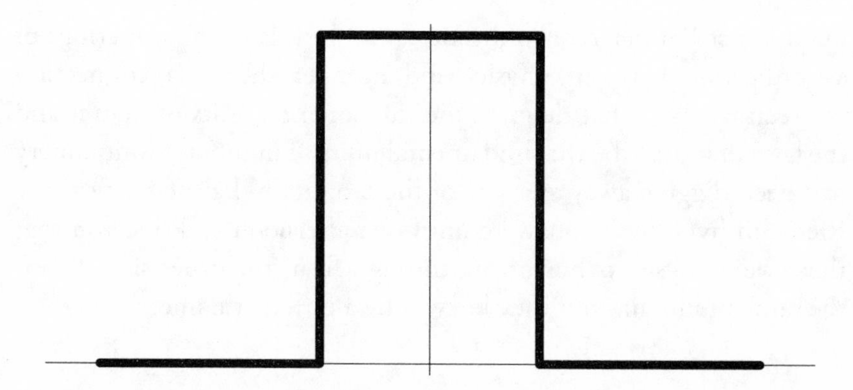

Figure 26. The boxcar function: zero everywhere, except for a short interval where it is exactly one.

contained the infinite. For Fourier analysis reveals the boxcar to be a wave built of the superposition of waves of all possible frequencies—it is the sound composed of all harmonics, all notes blended into one; the beam of light composed of all colors. Its pairing of the finite in time and the infinite in frequency shows that to constrain a sound in time necessitates an infinite orchestra. Translated to the wave phenomenon of light it shows that to constrain a color in space is to require a celestial palette. This mathematical object that represents each of these artistic paradigms is the same as that which seems to represent a picture of the connections embedded in the zeta zeros.

The Link of the Sinc

In and of itself, Montgomery's work was very exciting, marking a real breakthrough in the understanding of the structure of zeta zeros—assuming that they are where Riemann said they would be. But even more, it pointed to the possibility of a new direction of attack from which might emerge a proof of the Riemann hypothesis. In particular, Montgomery's work suggests that one road to a proof might be to find a way of producing a list of numbers that in turn will produce the telltale pair correlation evidenced by the zeta zeros. If so, then maybe a direct link between Riemann's zeta function and the newly discovered process can be found, and then maybe—just maybe—this will all lead to a proof of the Riemann hypothesis.

With the appearance of the sinc function we see the echoes of Riemann's first Fourier-zeta connection, and in this the shimmerings of a connection between physics and number theory, a connection between the laws that describe the fundamental units of matter and the laws that describe the fundamental units of number. Montgomery had been digging away at a side of the mountain that he knew as the Riemann hypothesis, unaware until one afternoon in Princeton that there were physicists burrowing through from the other side. It was the same mountain, but they knew it by a different name.

A RANDOM COLLISION WITH FREEMAN DYSON

Earlier on this fateful day Montgomery had given a lecture at the Institute to a small audience. He had discussed his recent work on the pair correlation of the zeros of Riemann's zeta function. Later that afternoon he arrived at tea with a friend who wanted to introduce him to the physicist Freeman Dyson, a professor in the School of Natural Sciences at the Institute.

While many a mathematician starts life as a physicist only to eventually flee the relative messiness of the physical sciences for the clean truths of mathematics, Dyson ran the reverse road. He was a mathematically gifted child prodigy born into a musically inclined family. His father, Sir George Dyson, eventually became director of the British Royal College of Music and his mother, Mildred Atkey Dyson, was a lawyer. Dyson's own first mathematical love was number theory, which he taught himself from Hardy and Wright's famous book *The Theory of Numbers.* Duly inspired, he enrolled at Cambridge University to study under Hardy.

World War II brought an abrupt halt to Dyson's studies. Only nineteen years old, he arrived for duty at RAF Bomber Command, sent there by the novelist C. P. Snow, who had been put in charge of wartime assignments for the technically trained and talented. Dyson traded the quiet contemplation of Platonic pure mathematics for the harsh wartime reality of the analysis of RAF bomber crew survival rates. RAF bomber formations and procedures were of ad hoc design at that point, infused by a great deal of folklore that at the time of Dyson's appointment seemed irrelevant and, what was even worse, life-threatening. After poring over the reams of mission data Dyson soon became the local expert on the statistics of RAF bomber collisions, and he was able to make recommendations leading to the adoption of new and improved (i.e., lifesaving) training regimens and scheduling protocols.

The role of science in the formulation of defense policy has played a large part in Dyson's professional life. In this he follows in the footsteps of his father, who was as well known for his authorship of the first British training manual on the throwing of hand grenades as he was for his music. Meanwhile, Dyson's academic father, Hardy, was an avowed pacifist. The combination of his upbringing and deep knowledge of nuclear physics has helped make Dyson one of the world's most eloquent and knowledgeable voices in the public discussion of nuclear weapons.

After the war, Dyson returned to finish his degree and found himself increasingly interested in physics. As the story goes, while leaving a physics lecture one day, he was speaking with a fellow Cambridge student, Harish-Chandra. Harish-Chandra, who was then an aspiring physicist, complained that the lecture revealed how messy a science physics was, and that because of this he was thinking of leaving physics for mathematics. Dyson is said to have remarked that he agreed about the messiness, but it was precisely for that reason that he was now thinking of leaving mathematics for physics. Years later the two would be reunited as colleagues on the Institute faculty, Harish-Chandra in the School of Mathematics, and Dyson in the School of Natural Sciences.

Much of Dyson's scientific fame derives from being the physicist who made mathematics out of *Feynman diagrams,* the intuitive and highly personal computational tricks and tools invented by the Nobel laureate Richard Feynman (1918–1988) to develop a working theory of *quantum electrodynamics* (QED). QED makes whole the wave-particle duality of the electron; it is able to unify phenomena that on the one hand suggest that an electron is a discrete particle and on the other that the electron is a continuous wave. Dyson's work set down the mathematical foundations of the subject, and by translating Feynman's art into mathematical formulas, enabled the rigorous statement of verifiable physical conjectures, among which was the successful prediction of the magnetic moment of the electron, first proposed by the physicist Julian Schwinger (b. 1918) and then by Feynman.

Dyson helped bring together the continuous and the discrete understandings of subatomic behavior. Similarly, by fusing his love of number theory with his expertise in creating the mathematical tools of physics, he would make the initial observation that would reinforce the connections between the discrete world of the integers and the continuous world of analysis, and thus galvanize research on the Riemann hypothesis.

As Dyson recalls it, he and Montgomery had crossed paths from time to time at the Institute nursery when picking up and dropping off their children. Nevertheless, they had not been formally introduced. In spite of Dyson's fame, Montgomery hadn't seen any purpose in meeting him. "What will we talk about?" is what Montgomery purportedly said when brought to tea. Nevertheless, Montgomery relented and upon being introduced, the amiable physicist asked the young number theorist about his work. Montgomery began to explain his recent results on the pair correlation, and Dyson stopped him short—"Did you get this?" he asked, writing down a particular mathematical formula. Montgomery almost fell over in surprise: Dyson had written down the sinc-infused pair correlation function.

Dyson had the right answer, but until that moment he had associated this formula with understanding a phenomenon that seemed completely unrelated to the primes and the Riemann hypothesis. In a flash he had drawn the analogy between the sinc-described structured repulsion of the zeta zeros and a similar tension seemingly exhibited by the different levels of energy displayed by atomic nuclei. Whereas Montgomery had traveled a number theorist's road to a "prime picture" of the pair correlation, Dyson had arrived at this formula through the study of these energy levels in the mathematics of ***matrices.*** This connection is the source of most of the current excitement surrounding the Riemann hypothesis, and it's time to see just what all the fuss is about.

MATRICES: THE MATHEMATICIAN'S SPREADSHEETS

If numbers represent our first tentative attempts to organize the world, then one way in which we continue to express our human penchant for classification is the further organization of this basic material into the tables upon tables of numbers that accompany us through life. Sports-minded children cut their teeth on the numerical tables of statistics. Baseball fans scan the neatly displayed data, indexed by years and statistics, the columns of batting averages and runs batted in, home runs, doubles, triples, walks, and strikeouts. The young basketball zealot ponders lists of free throws, points per game, rebounds per game, and assists. There is magic in the distillation of sports heroes to lists within lists; facts to be argued over in the school yard and on the walk home; numbers within numbers whose precise memorization is a quiet, unconscious achievement and a source of parental wonder and pride.

As we grow older, sports take a backseat (or maybe the passenger seat) while commerce and new numerical tables enter our lives. Batting averages give way to the Dow Jones, and baseball cards are replaced by the business pages and their tables upon tables of financial data. Rows are indexed by company names, columns by various attributes: the fifty-two-week high, the fifty-two-week low, yesterday's closing price, volatility, etc. Bingo players stare intently at the tables formed by their cards, a seemingly random array of integers that may hold the key to some quick cash. Actuaries scan life contingency tables hoping to unravel the statistical secrets of expected life spans. More and more of us open spreadsheets to create budgets or do business through the use of these convincing and orderly numerical displays.

In each of these examples we see a ***matrix***. In its simplest form, this is a rectangular checkerboard-like grid filled with numbers, one to a square. To a mathematician, these numerical arrays have as much in common with the movie *The Matrix* as with the accountant's bookkeeping tool, for the matrix of the mathematician is all about transformation. Dyson was concerned with the matrices that made sense of the transformations in the atom. To understand this, however, let's look at a more down-to-earth example.

A Codification of Transformation

It's autumn in New York City. Imagine that you awake in your apartment up high in the sky. The sun is sneaking in through the blinds, a sign of a beautiful, bright, crisp fall morning. You loll around until nuzzled into consciousness by your big friendly golden retriever Digger, reminding you (how could you forget?) that it is time to head to Central Park. You stumble out of bed, wash quickly, dress, grab your coat, and head to the door. Before leaving you make sure that you have with you the new designer "Operator" brand sunglasses that you bought yesterday on Madison Avenue. They are sleekly styled and all the rage, marketed with the unusual feature that they come with a variety of lenses, each specifically tailored to a particular environment. You put the glasses and their assortment of lenses in your pocket and, leash in hand, leave for the park.

You enter Central Park and eventually reach the oval surrounding the expanse that is the Great Lawn. Here, away from the trees, you find the sun in your eyes, so you reach into your pocket to put on your new shades. Among the selection of lenses is one labeled "NYC." You pop these in, put on your shades, and are immediately soothed by a world whose colors are changed only in intensity, now lessened to a softer version of their former selves. This is good enough for walking the dog, but the other lenses pique your curiosity. There are the "Everest" lenses, which must have been designed to mitigate the bright light of high-altitude sunshine. With some more rummaging around you come upon the "Wonderland" lenses. These need to be experienced.

You pop them in, look about, and become aware of the transformation undergone by the park, literally, right before your very eyes. The variegated greens of the grassy field have been transformed to shades of purple. The red-and-blue leash in your hand is now a similarly patterned collage of yellow and orange, while your dear old dog, ordinarily the color of the fallen leaves all about you, has been transformed into a strange canine alien. His usual coat of nuanced reddish-brown is now a weird, unearthly combination of yellows, purples, and oranges that at least in relative tone seems to mirror the color composition of his usual appearance.

Stunned, you remove your sunglasses. What are these "Operator"-induced effects? To describe these visual transformations we use a matrix.

Any color can be represented as a mixture of varying amounts of red, green, and blue, ranging from pure black, realized (or unrealized) as the absence of all three primary colors, to the pure white obtained as the combination of all three, each at full intensity. In this way, any color is naturally, and mathematically, uniquely identified by a *3-vector*, a list of three (nonnegative) numbers whose values indicate the individual intensities of red, green, and blue. With this sort of dictionary, pure red is translated to (1, 0, 0); a dull green might be (0, 1/2, 0); and (0, 0, 5) represents a deep blue. Deep purple might be (2, 3/10, 7), a mixture of dark red and darker blue, with a dash of green for texture.

In this scheme, the transformative effects of your "Operator" sunglasses have their own mathematical translation. We begin by describing the effects on the primary colors red, green, and blue. The NYC lenses seemed to do little more than reduce the intensity of every color. A measurement might show that each of these colors was reduced in scale to one-tenth its original intensity: the pure red of (1, 0, 0) is transformed to (1/10, 0, 0), with analogous transformations to (0, 1/10, 0) and (0, 0, 1/10) for pure green and pure blue respectively. This could all be summarized in a table or matrix of three columns, the first of which is made up of the transformed red, the second of the transformed blue, and the third of the transformed green. This is shown in Figure 27.

In particular, Figure 27 shows a *three-by-three* (for three rows and three columns) *diagonal matrix*, so named because the only nonzero entries occur along its diagonal.

	Red	Blue	Green
Red	1/10	0	0
Blue	0	1/10	0
Green	0	0	1/10

Figure 27. Three-by-three diagonal matrix (i.e., three rows and three columns) that describes the effect of the NYC lenses. The first column records the effect on pure red, the second on pure blue, and the third on pure green. In each case, only the intensity of the individual color is affected. Thus the first column records the fact that pure red is transformed to red of one-tenth the intensity, with no additional contributions of either blue or green; hence the entries of zero in the other two rows of the first column.

	Red	Blue	Green
Red	3	1/2	7
Blue	1/2	1	9
Green	7	9	2/3

Figure 28. Three-by-three symmetric matrix (i.e., three rows and three columns) describing the effect of the Wonderland lenses. The first column records their effect on pure red, the second on pure blue, and the third on pure green. For example, the matrix here indicates that when you wear the Wonderland lenses, red will appear as a color that in fact has a fair amount of blue in it (as indicated by the red-blue entry, which is in the first column and third row. Green will turn into a color that has relatively little red in it (as indicated by the green-red entry, which is in the first row, third column).

Now we do the same for the Wonderland lenses. Perhaps the pure red is transformed to a color represented by (3 1/2, 7), pure green is turned into (1/2, 1, 9), and pure blue becomes (7, 9, 2/3).

From these representative transformations we again make a table, as shown in Figure 28. The first column records the transformation of red, the second the transformation of blue, and the third the transformation of green. This encoding of the effects of the Wonderland lenses again results in a three-by-three matrix. The fact that the entry in any given row and column in the matrix in Figure 28 is equal to the entry in which the row and column are reversed characterizes this as a *symmetric matrix.*

Under what conditions would the matrix of Figure 28 contain enough information to describe the transformation effected on any color? Our experience in Central Park suggests that scaled intensities of color are transformed in a predictable way: for example, that a shade of red which is one-tenth as deep as our exemplar red is turned into a color represented by the vector whose intensities are (1/10 × 3, 1/10 × 1/2, 1/10 × 7), which is (3/10, 1/20, 7/10), or one-tenth of those that represent the transformation of red. Similarly, a shade of red twice as deep as the paradigm would appear as that color with each intensity doubled.

Also, we notice that the effect of the lenses seems *additive.* For example, consider our experience of the transformation of the color purple. Most of us know that purple is generally the combination of red and blue. In our new vector representation of color, then, the effect of our glasses on purple is simply the vector represented by the sum of its independent transformations of red and blue, a color represented by (3 + 1/2, 1/2 + 1, 7 + 9), the entry-by-entry sum of the transformed red and blue.

The assumption that the Wonderland lenses respect the effects of scale and additivity describes it as a *linear transformation* of the three-dimensional "color space." In this situation the defining matrix contains all the information necessary to predict the visual effect experienced in looking at any swatch of color while wearing the "Operator" sunglasses with the corresponding lens. By expressing any color as the sum of some appropriately scaled amounts of red, green, and blue, we now see that it will appear as the sum of the equally scaled amounts of the transformed red, green, and blue, represented by the columns of the matrix.

What if the Everest lenses were placed over the Wonderland lenses? While we could of course try this out and write down the new matrix that this experiment would yield, there is in fact a way of multiplying matrices that would permit us to perform a simple calculation using the matrix of the Wonderland lenses with the matrix of the Everest lenses to obtain the result of wearing the latter in front of the former. And of course we could keep going, and, by performing a sequence of such operations, predict the effect of wearing any combination of lenses.*

Fittingly, the introduction of this systematic codification is usually attributed to the British mathematician and lawyer Arthur Cayley (1821–1895). Cayley was among the most prolific of mathematicians,

*Matrices are in this formal way much like the integers, capable of combination by a means of addition as well as multiplication (mathematicians call such a structure a *ring*), but of a type that has a few surprises. Multiplication of integers is ***commutative***—meaning that the product 2 × 3 gives the same result as 3 × 2—but in multiplying matrices the order matters. It is possible for different results to be obtained, depending on which one is on the right-hand side of the product and which is on the left. In terms of our Operator sunglasses, we might get different views of the world depending on the order in which we insert the lenses. In this way, matrix multiplication is inherently ***noncommutative.*** Stranger still, it can be possible to multiply two nonzero matrices and get zero. But these are good surprises, indicating a rich and fascinating subject of study.

writing more than 700 mathematical papers, a good third of which were written while he ran a highly successful legal practice. He is one of the great *algebraists,* the mathematical lawmakers who formulate the rules of symbolic manipulation. Akin to the axiomatic approach to geometry, algebra often proceeds by setting down a few simple rules that dictate how symbols may be conjoined and proceeds to investigate the ramifications. Even the most simple collection of laws can generate an incredible diversity of unexpected behaviors. This is a phenomenon probably quite familiar to the head of a busy legal practice.

Cayley introduced matrices to study certain transformations of the Cartesian plane. They became the spreadsheet of the geometer, providing a mechanism to encode the pushes and pulls, twists and turns that can distort an object's shape in any dimension. The linear transformations of the two-dimensional plane are but the simplest case. We could of course go on to the three dimensions of spatial experience or the four dimensions of space-time. But three- and four-dimensional examples are small potatoes. The matrices that arise in modeling the airflow over an airplane wing or the stresses on a suspension bridge may be 10,000 by 10,000, a number of dimensions reflecting the minute scale of analysis and the number of parameters necessary for their understanding. Even these examples pale in comparison with the matrices needed to describe the pushes and pulls within the world of the atom. Although this is an infinitesimal world, its description requires infinite-dimensional matrices. These are the ***Hamiltonians,*** the matrices behind Dyson's inspired guess, and we now turn to their explanation.

Beware the Hamiltonian

Hamiltonians are named for the Irish mathematician William Rowan Hamilton (1805–1865). Hamilton was another child prodigy, raised by an uncle who was a linguist; it is said that Hamilton could read Latin, Hebrew, and Greek by the age of five, and that he spoke at least six eastern languages by age ten. While still an undergraduate at Dublin's Trinity College, he was named royal astronomer of Ireland and professor of astronomy. At thirty he was knighted for his scientific achievements.

Dyson's unconscious arrival at the gates of the Riemann hypothesis had come after years of study devoted to understanding the inner workings of the atomic nucleus as described by the Hamiltonian matrices. At this time the Nobel Prize–winning achievements of Feynman and Schwinger (who shared the 1965 prize for the development of QED with the Japanese physicist S.-I. Tomonaga) had created a working atomic theory, but it required a simplifying assumption in which the nucleus was treated as a positively charged point mass. The next step in understanding the atom would need to account for the nucleus as it really was: a miasma of complicated interactions among the nucleons, the subatomic particles that occupy the nucleus. Nucleons come in two flavors—neutrons and protons—and they coexist in the nucleus, held together by a mysterious *strong force,* which needs to fight against the electromagnetic repulsion felt between all like charged protons. This tightly wound bundle of energy is perpetually bathed in and influenced by the surrounding electrons. In toto, these subatomic particles engage in a subtle and complicated dance of repulsion and attraction, one whose complexity grows dramatically as the number of particulate dancers increases. The Hamiltonian for the nucleus must account for all this. Moreover, it must also reflect the fact the nucleus is described not by classical (i.e., Newtonian) physics but rather by the physics of ***quantum mechanics.***

Classical physics is enough for plotting the trajectory of a billiard ball struck by a pool cue or describing the forces keeping an airplane in flight or a car glued to the road, but inside the atom this model falls apart. The failure of the classical model to agree with known experimental facts led to the development of quantum mechanics. In the quantum mechanical model of the atomic or nuclear world, an exact knowledge of position and momentum are replaced by a probabilistic certitude, embodied in a mathematical function called a ***wave function,*** which can tell you only the probability of a particular configuration at any time and place. Within this probabilistic outlook is embedded an uncertainty that is not part of the world of classical physics. Even the ability to simultaneously measure exactly an object's velocity and position is lost.

Whereas the deterministic world gives us models of two-, three-, four-, and other finite-dimensional systems, the probabilities that are encoded in the wave function occur in a mathematical land of infinite dimensions called a ***Hilbert space,*** another mathematical discovery named for David Hilbert (see Chapter 9). The Hamiltonian matrix is a spreadsheet with an infinity of rows and columns that keeps track of the quantum mechanical system. It tells the tale of the uncertain possibilities that could befall an energetic traveler in the nucleus.

Dyson would connect Hamiltonians and zeta zeros through the *energy levels* encoded in the Hamiltonian. If the nucleus is a tempest of fundamental forces, then the energy level measures the tossing and turnings of a specific state of these submicroscopic seas. A classical system can assume a continuum of energetic states (think of a stone raised ever higher and higher off the ground, steadily gaining potential energy); but one irony of quantum mechanics is that in all its uncertainty, only a discrete (albeit infinite) collection of possible energies can ever be observed in a quantum mechanical system.

The energy levels of an atom or atomic nucleus are often revealed through *spectroscopy.* In these experiments the atom or its nucleus is bombarded by radiation of a particular energy, which is absorbed and then released, like the excitation of a (perfect) tuning fork by a perfectly executed whack which tingles its tines. Whereas the tuning fork will then resonate at a single intrinsic frequency, which you hear as its pitch, the atom is effectively an infinite, countable number of tuning forks that are set into motion by the radiation. The subsequent atomic chorus is realized in the appearance of ***spectral*** lines that indicate the energy levels of radiation emitted by the atom as it relaxes back to equilibrium. Atomic spectroscopy requires excitation of the electrons, and much of this we witness in the visible range of the spectrum bounded on the low end by infrared radiation and on the upper end by ultraviolet, which bracket the hero of spectroscopy: ROY G. BIV (a mnemonic for red, orange, yellow, green, blue, indigo, violet). This is but a more energetic version of the various instances of phosphorescence or fluorescence that we often see in a local science museum or a well-appointed shop selling rocks and minerals.

The totality of energy levels in this range is the atomic ***spectrum,*** and the irregularly spaced array of lines are an atomic fingerprint, a unique identifier of the elements akin to the whirls and whorls of each of our own hands. Similarly, the various atomic nuclei have their own fingerprints. These occur at much higher energy levels, typically around the range of X-ray radiation. In these sequences of spectral lines and energy levels for heavy nuclei, those with more than the simple proton nucleus within hydrogen, we see the hints of a Riemannian connection, a tantalizing specter of the markings of the zeta zeros on the critical line.

In the case of a relatively simple atom like hydrogen the associated Hamiltonian can be written directly. With its single proton and electron, the hydrogen atom comprises a tiny subatomic universe, simple enough to permit the calculation of its concomitant atomic and nuclear energy levels.

But what about more complicated nuclei? With the addition of even one more nucleon the specific dynamics begin to defy description. How can we write down the Hamiltonian of this system? How to understand the energy levels of the system? That is the challenge which Dyson and others took up, and in order to begin to understand the situation they turned to the tools of statistics and sought insight through the consideration of ***random matrices.*** This is the investigation that leads to the Riemann hypothesis.

Statistical Knowledge: Understanding through Randomness

The motivation for this chance-inspired approach can be found in the methodology used to understand other kinds of tremendously complicated phenomena, such as the behavior of a gas interpreted as the interactions among the 10^{23} molecules bouncing within a heated balloon or the millions of tiny muscles that power our hearts. A naive approach would be to try to scale up the analysis of a simple single basic interaction to the level of the collective phenomenon. In the case of our heated balloon, this would mean taking the classical analysis of a single perfectly caroming billiard ball and hoping to use that understanding to shed light on the confusion of the interactions of a huge collection of balls on the table. The sheer numbers makes this impossible. Instead, years ago, the physicist Ludwig Boltzmann (1844–1906) suggested that we look for large-scale statistical behavior. Ask instead what is happening on average, and the amazing thing is that, often, this is very close to what we observe happens individually.

This is the arena of *statistical physics.* Because of the analytical difficulties encountered in trying to understand the dynamics governing the interactions of several or many subnuclear particles, Eugene Wigner (1902–1995) of Princeton University, a Nobel laureate, proposed treating a complex nucleus as a "black box" in which a large number of particles are interacting according to unknown laws, ostensibly described by some unknown Hamiltonian. Wigner is famous for (among many other things) coining the phrase "the unreasonable effectiveness of mathematics," to describe the mysterious manner in which mathematics—even those parts of mathematics that were initially dreamed up as mere intellectual fancy—so often turns out to be so well suited to describing much of the physical world. Dyson and others came to believe that, yes, these complex heavy nuclei are too complicated to understand in the same way we can understand the hydrogen atom. So instead we'll try to understand the *ensemble* behavior—i.e., the range or distribution of behaviors that will appear over a variety of possible nuclear situations.

For example, a statistical-mechanics approach to understanding a single molecule in a balloon filled with gas calculates the average behavior of such a particle. This is equivalent to predicting the behavior of an individual particle chosen randomly from an ensemble of particles that make up a system described by a particular temperature and pressure. An analogous and surely more familiar example would be to consider an experiment in which 1,000 fair coins are tossed simultaneously. The outcome of any particular coin is impossible to predict accurately—it is equally likely to be heads or tails—but we have a fairly good understanding of the ensemble behavior. We are confident that close to 500 of the coins will come up heads (and the remainder tails), and we have a good understanding of how the collective outcome might vary from this most likely event. Applying the same philosophy to the problem of understanding heavy nuclei, we'll envision the dynamics of such a system as being but one description among an ensemble of possible descriptions, an infinite family of transformations of functions of time and space, necessarily circumscribed by certain considerations of symmetry and regularity as befits the allowable transformations of the subatomic world. In doing so, we hope we now have posed a problem that is tractable as well as informative.

This was the approach that Wigner, Dyson, and many others (especially the physicists Michel Gaudin and Madan Lal Mehta) took to investigate heavy nuclei. They studied the properties of these ensembles of models of the nuclei as the ensembles ranged over a very loose set of constraints. Mathematically, this meant that they considered the structure of the Hamiltonians representing these nuclear environments. They began to study the statistics of these matrices, which necessarily satisfied a ***Hermitian*** symmetry condition. This sort of symmetry, named for that old Stieltjes supporter Charles Hermite (see Chapters 6 and 7), embodies the requirement that over time, the transformation encoded in the Hamiltonian neither stretches nor squashes space.

Finally, as these Hamiltonians are the transformations of some infinite-dimensional space, these matrices would need to be studied asymptotically—i.e., analyzed at the limit, as the dimension of the space that is being transformed grew beyond any preassigned bound. By doing this they would begin to understand something of the structure of a random Hamiltonian, and ultimately something of the distribution of energy levels of "typical" heavy nuclei, among which should be the heavy nuclei of the elements of our world.

What Dyson intuited was that the energy levels, as determined by a random matrix, must have some of the same properties as the zeta zeros. But energy levels are physical quantities which we hope to measure with machines. The matrices that embody these Hamiltonians yield predictions of these physically observable quantities, provided by their ***eigenvalues.*** This is the last link of the bridge from Hamiltonians to the zeta zeros.

The "ei" of the Storm: Eigenvectors and Eigenvalues

Like the atomic spectrum which they translate into mathematics, eigenvalues and their associated ***eigenvectors*** serve as the fingerprint of a matrix. For example, although—as we saw through the lenses of our "Operator" sunglasses—the way in which matrices (lenses) transform vectors (colors) can in general be quite surprising, there are in fact various inputs for which the effects simplify, like the goings-on in the eye of a hurricane. These inputs are the eigenvectors of the matrix. Harking back to our sunglasses, the eigenvectors of a particular lens would be those swatches of color which appear to change only in intensity, not in hue. The intensity of this effect of dulling or heightening of color is the associated eigenvalue.

Our three-by-three matrices that correspond to our sunglass lenses have only three eigenvectors, but the Hamiltonians transforming an infinite-dimensional space have an infinity of eigenvectors. They correspond to an infinite discrete set of behaviors, whose superpositions describe all possible subatomic configurations and whose energies are given by the corresponding eigenvalues. Arbitrary wave phenomena can be described as superpositions (i.e., infinite summations) of appropriately modulated basic sinusoidal waves; Riemann found a description of the number of primes of a given magnitude in terms of a summation of analogous zeta-zero-indexed oscillatory phenomena. Similarly, any wave function in the atom can be expressed as an infinite superposition of these eigenvectors. In this way the eigenvectors are the fundamental building blocks for describing the possibilities for life in the atom.

Implicit in this discussion is the assumption (as in the example of the lenses) of Hermitian symmetry. Realized in terms of matrices, this implies a certain symmetry of pattern in the numerical entries, which are now drawn from complex numbers. These nuclear Hamiltonian matrices have a symmetry analogous to that of the matrices of the Wonderland lens: rather than corresponding entries on opposite sides of the northwest-southeast diagonal being equal, they are instead *complex conjugates* of one another: these associated entries have the same real part, but their imaginary parts are the negatives of one

-1	8	$-4i$
8	3	$-2+5i$
$4i$	$-2-5i$	18

Figure 29. Example of a three-by-three Hermitian matrix. The symmetry is such that corresponding entries on either side of the diagonal entries (e.g., the entries in row two, column three and row three, column two are "corresponding" in this sense) are related by having the same real part (equal to -2), but the imaginary parts are the negatives of each other (equal to $5i$ and $-5i$ respectively).

another. This also implies that the entries on the diagonal must be real numbers. An example of a three-by-three matrix with Hermitian symmetry is shown in Figure 29. This symmetry guarantees that the eigenvalues of the Hamiltonian, and thus, the energy levels of the corresponding system, are real numbers.

The random connection

Stieltjes had inspired a search for a road to Riemann through randomness, believing that he had found a proof of the Riemann hypothesis in the seemingly less than random walk of the Möbius function. Dyson's path to the primes was his realization that the zeta zeros might very well be like the eigenvalues found in one of the myriad of matrices that are Hermitian Hamiltonians. The magical coincidence, or what Wigner might have called the "unreasonable effectiveness of physics," is that these numbers, which are on average the energy levels of sufficiently complicated nuclei, bear an uncanny resemblance to the zeros of Riemann's zeta function. In other words, the numbers that describe the primal waves of the prime counting function are much like numbers that describe the energy levels of the fundamental wave functions of an "average" heavy nucleus. This is the content of Dyson's remark in the tearoom at the Institute.

The connection is indeed astounding, but as Dyson recalls, he had certainly been primed for it. In the early part of the twentieth century, Hilbert and George Pólya (1887–1985) had independently proposed an approach to settling the Riemann hypothesis by finding an ***operator*** (i.e., matrix) whose eigenvalues were the zeros of Riemann's zeta function.

We've already met Hilbert. Pólya was also a remarkable figure in mathematical history. He was among the brilliant Hungarian scientists forced to leave their homeland in the 1930s. He spent most of his academic career at Stanford University. Beyond his mathematical work he is perhaps best known for his investigations into the process of problem solving. His two-volume work *Induction and Analogy in Mathematics* is a compendium of mathematical problems, their solutions, and expositions of the process of finding these solutions. His popular account of the thought process used in solving mathematics problems is *How to Solve It,* which has sold over 1 million copies to date.

Pólya had tremendous intellectual and mathematical breadth. Among his many achievements are some of the first rigorous results in the theory of random walks that had inspired the investigation of Brownian motion. Pólya developed an interest in this area while wandering in the woods one day outside Zurich. Over the course of his walk, he kept bumping into his roommate, who was spending the day strolling with his fiancée. These frequent meetings greatly embarrassed Pólya, and that evening he decided to see if mathematics could help show why they had kept happening. Pólya analyzed a simplified version of the situation, the "drunkard's walk," in which an inebriated protagonist navigates a gridwork of streets by choosing randomly among directions available at each corner: east, west, north, south. Under this model Pólya was able to show that with only rare exceptions, these random walks will visit every position in the grid infinitely often, a fact succinctly summarized as a "random walk in two dimensions is recurrent." An analogous statement is also true in one dimension (visualized as the same drunkard careening up and down a single avenue), but it is a wondrous fact of mathematics that in three dimensions (realized as random clamberings along an infinite jungle gym) recurrence is no longer true. In three (or more) dimensions the drunkard may never find his way home.

Of particular interest for our story is Pólya's work on the zeros of *integral operators.* One particular example is the Fourier transform, a prismlike process which takes as input any sort of wave phenomenon and breaks it into its fundamental components. If a mathematical function takes numbers in to spit new numbers out, an integral operator kicks the level of abstraction up a notch. In the case of the Fourier transform, the procedure takes in a function of time or space (such as the amplitude, or volume, of a wave at a given time) and returns the new function that measures how much of each basic frequency is contained in the wave. Integral operators share many of the properties of matrices, but like the Hamiltonian these are matrices of an infinite number of dimensions. Like most mathematical functions, these operators can also vanish for certain inputs, and the inputs that cause this are their zeros.

In the early twentieth century, Pólya had proved a variety of facts about the zeros of these integral operators, all directed toward characterizing Riemann's zeta function. Late in his life, Pólya recalled:

> *I spent two years in Göttingen ending around the begin [sic] of 1914. I tried to learn analytic number theory from Landau. He asked me one day: "You know some physics. Do you know a physical reason that the Riemann hypothesis should be true." This would be the case, I answered, if the nontrivial zeros . . . were so connected with the physical problem that the Riemann hypothesis would be equivalent to the fact that all the eigenvalues of the physical problem are real.* *

At around the same time that Pólya was in Göttingen, Hilbert had been conducting a similar line of research in his work related to the development of Hilbert spaces, those mathematical structures that give a home to the wave functions of the atom. Physicists and mathematicians were then trying to reconcile two mathematical approaches to quantum mechanics: one, due to Heisenberg, that relied on matrices; and another, due to Erwin Schrödinger (1887–1961), that focused on a central differential equation, the wave equation, which gave rise to the idea of a wave function. While an exact reference for Hilbert's work in this regard is still unknown, there is one intriguing bit of related history to be found in Constance Reid's outstanding biography *Hilbert.* Reid reports that Hilbert had predicted the equivalence of these approaches to quantum mechanics and that his evidence came from a wealth of experience in which matrices always accompanied the solutions of certain sorts of differential equations. Taken together, the two approaches of Hilbert and Pólya speak to this idea of finding an infinite-dimensional matrix whose eigenvalues could correspond to the zeros of Riemann's zeta function. Although neither Pólya nor Hilbert ever published (or publicized) these remarks, they became part of the folklore of number theory. So was born the *Pólya-Hilbert approach* to the resolution of the Riemann hypothesis: the search for a bridge between eigenvalues and zeta zeros.

*Courtesy of A. Odlyzko. See http://www.dtc.umn.edu/~odlyzko/polya.

Someone like Dyson, reared as a number theorist, was well aware of the Pólya-Hilbert approach. Just as a random matrix might incorporate the complexities of the inner workings of the atom, perhaps it might shed light on a Holy Grail of matrices: the operator whose eigenvalues were the zeros of Riemann's zeta function. In the random walk that often represents intellectual progress, Dyson's chance encounter with Montgomery revealed the link that would bring the mathematical and Riemann-wrangling world back to Pólya and Hilbert.

THE DYSON-ODLYZKO-MONTGOMERY LAW

The announcement of Montgomery's results and Dyson's connection inspired Andrew Odlyzko (b. 1949) to start some computations. Montgomery had needed to make some technical and unproved assumptions about the zeta zeros and Riemann's zeta function in order to derive his beautiful formula for the pair correlation. Odlyzko wanted to see just what the naked data would really show. Since Dyson and Montgomery had derived asymptotic results, Odlyzko knew that in order to have any hope at all of replicating this phenomenon he'd have to compute using a huge number of zeta zeros in the stratosphere of the critical strip. This is made a little easier by the fact that the zeta zeros occur closer together as we move farther and farther up the critical line.

At first glance the thin, bespectacled, soft-spoken Odlyzko hardly seems the type to be engaging in computations at the limits of current computational power, but his manner belies an intensity and a wry sense of humor that are just what it takes to be working in this realm, where disappointment can often follow weeks of computation.

Like many a mathematician, Odlyzko entered scientific life with an attraction to the physical and life sciences, but was put off by the amount of time spent performing menial and unscientific laboratory tasks as well as a lack of "rigor" in the way in which mathematics was used. Thus he decided to concentrate on the field of mathematics, in which he felt as though he could "see more directly" the new things that he was contributing.

Odlyzko is now a professor of mathematics at the University of

Minnesota, and these days is considered the master of zeta zero calculations. He is also famous for joint work with the Dutch mathematician Herman te Riele that provided a *dis*proof of the Mertens conjecture, thereby putting a damper on a Stieltjes-like approach to the Riemann hypothesis through the meanderings of the Möbius function.

In the 1980s, when Odlyzko took up the challenge of the zeta zeros, he was a research scientist at Bell Laboratories. The culmination of this work was a 163-page paper, never published but circulated like a piece of mathematical samizdat, detailing a wide range of zeta zero calculations. In a set of calculations performed at what were then the edges of Riemann's zeta zero universe, Odlyzko computed the 10^{20} − 30,769,710th zero through the 10^{20} + 144,818,015th zero (this is more than 170 million zeros). He then proceeded to compute the pair correlation arising from the first 80 million zeros occurring beyond the 10^{20}th (that's the one-hundred-billion-billionth) zero. Odlyzko was the first to compute zeros in the search for structure rather than a counterexample, thus revealing a true paradigm shift in the thought process related to computational approaches toward the Riemann

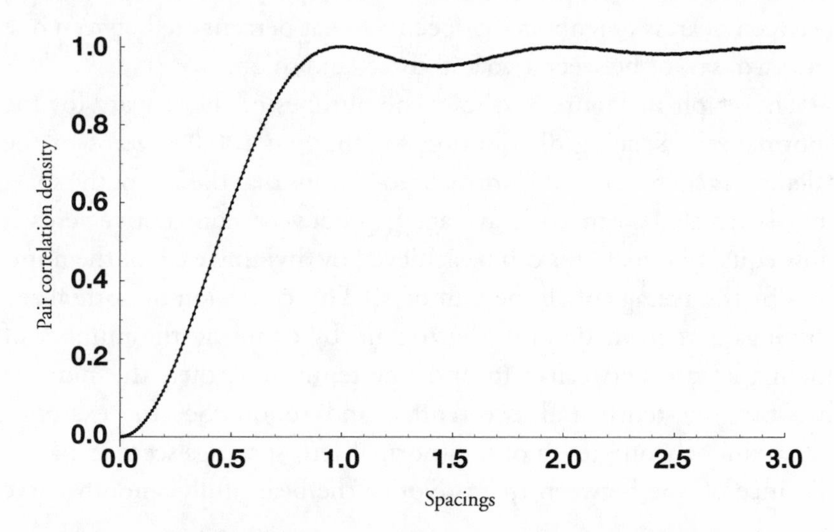

Figure 30. The pair correlation for 1 billion zeta zeros, starting with the zeta zero that is number 10^{23} + 985,532,550 in the list. It is compared with Montgomery's smooth sinc-derived pair correlation. (Data provided by A. Odlyzko, graph by P. Kostelec.)

hypothesis. His calculations showed a virtually perfect agreement with Montgomery's theoretical prediction.

The coupling of Odlyzko's calculations with the investigations of Montgomery and Dyson naturally gives rise to a new conjecture: that the zeros of the zeta function and the eigenvalues of a random Hermitian matrix have exactly the same pair correlation. This is often called the *Montgomery-Odlyzko law,* but more properly, perhaps it should be called the ***Dyson-Montgomery-Odlyzko law.***

The pair correlation is just one statistical measure of the spacings between the zeros—or eigenvalues—and it is a coarse measure at that. By considering the distribution among all possible differences between numbers in a list, the pair correlation effectively lumps all the spacings together. A more detailed analysis is the consideration of the nearest-neighbor ***spacings,*** the distribution given by keeping track of the distance between successive zeta zeros. Notice that if you know this, then you know the "gap" between any two (even nonsuccessive) zeros. It is simply the sum of the gaps between the succession of nearest neighbors that fall between the two.

So, following our collective statistical nose, we consider the ***spacing distribution:*** i.e., what are the various proportions of spacings between nearest neighbors that occur? What percent are between one and two, say, or between two and three, and so on?

The graph in Figure 31 shows the outlines of the density for the "normalized" spacing distribution for the first 100,000 zeros of the Riemann zeta function.* "Normalized" means that the size of the spacings is rescaled so that the average gap between consecutive zeros is now equal to one. (This can be achieved by dividing each of the numbers by the average of all the numbers.) This collection of normalized spacings is then made into a histogram by counting the number of them that are between zero and one-tenth, and then the number between one-tenth and two-tenths, and so on. So, for example, approximately one-tenth of the (normalized) spacings seem to have a distance of one between them. Notice the beautifully smooth curve

*Following tradition, we speak of the "distribution" of a collection of numbers, but show the graphs of the density. For example, the "bell curve" is the shape of the density for the normal distribution.

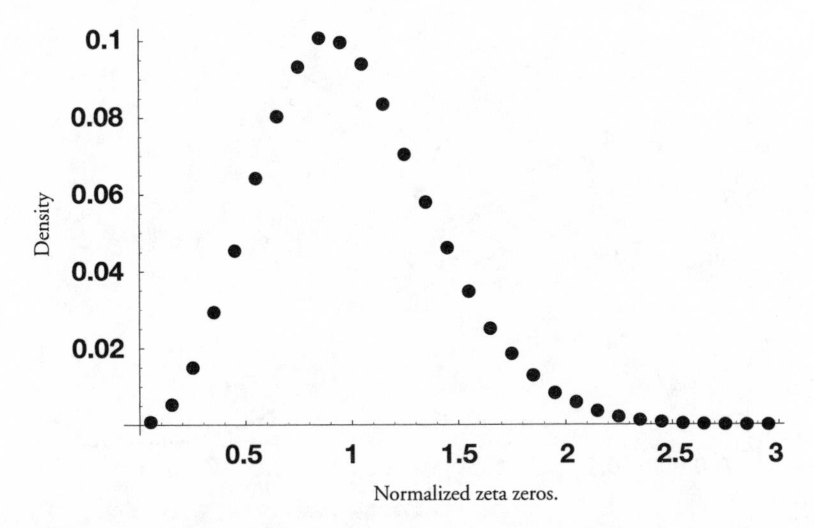

Figure 31. Histogram of the "normalized" spacing distribution for the first 100,000 zeta zeros.

upon which the dots seem to fall. Even in the first 100,000 zeta zeros we can see structure emerging.

We can compute the same thing for the ordered eigenvalues of a Hermitian matrix. Moreover, as in the case of the pair correlation, we can compute the expected distribution of the spacings for a random matrix. It turns out that asymptotically, for a random matrix, this distribution is known: it is the smooth, nearly bell-shape ***Gaudin distribution,*** named for its discoverer, the random-matrix researcher M. Gaudin. It is also known as the *GUE spacing distribution,* where GUE stands for Gaussian unitary ensemble, reflecting a particular "unitary" symmetry satisfied by these random Hermitian matrices.

Once again Odlyzko compares the spacing distribution of the zeta zeros with that of a random Hermitian matrix. In his landmark paper, using the first 80 million zeta zeros past the 10^{20}th, he performs the calculations of the spacing distribution. Comparison with the Gaudin distribution shown in Figure 32 shows that the agreement is spot-on.

Both the pair correlation and the spacing distribution of the zeta zeros reveal a seemingly undeniable equality to the analogous distributions computed for the eigenvalues of random Hermitian matrices. Odlyzko has continued (in his spare time) to compute zeros of Riemann's zeta function, and the agreement between the spacing distri-

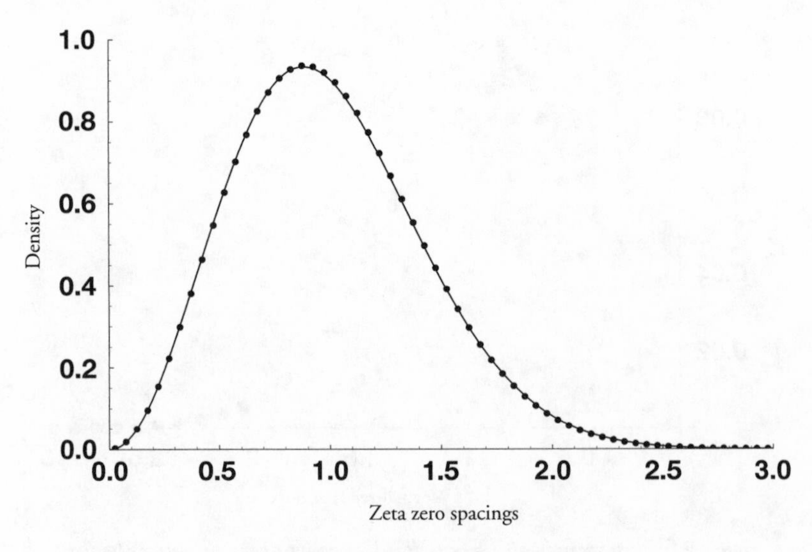

Figure 32. Density for the (normalized) nearest-neighbor spacing distribution for 1 billion zeta zeros, starting with the zeta zero that is number 10^{23} + 985,532,550 in the list. It is compared with the smooth bell-like density for the Gaudin distribution. Note the virtually perfect agreement. (Data provided by A. Odlyzko, graph by P. Kostelec.)

bution of the zeros with the spacing distribution for random matrices keeps getting better and better. Odlyzko has said that "there is a beauty to that connection" uncovered by his computations, a beauty which he likens to the elegance of the Brooklyn Bridge and the feeling it stirs in him, a "resonance with our inner soul."

As computers get faster and faster, more and more zeros are computed and better and better agreement is seen. Surprisingly, Odlyzko believes that the increased speed of today's computers is less responsible for the execution of the more extensive calculations than the brainpower that has gone into improving the computational techniques ever since people have begun to put pen to paper in order to perform zeta zero calculations. By his estimation, improved technology is responsible for a 10-billion-fold improvement in the efficiency of computing zeta zeros, but the algorithmic advances provide on the order of 10,000 billion times the original efficiency. Even in the world of the computer, we are given a John Henry–like lesson, reminded of the advantages of human versus engineered brainpower.

The current favorite fast algorithms make great use of the mathematics of digital signal processing, including the famous fast Fourier transform. So we have come full circle, using the technique first discovered by Gauss in order to discern the hidden orbits of asteroids to now uncover the patterns hidden in Gauss's original calculations of the density of primes.

A road to Riemann through randomness?

The mathematician Peter Sarnak of Princeton calls Odlyzko's computations "one of the most important mathematical achievements of the twentieth century." They provide empirical substance to the dreams of Pólya and Hilbert and have redirected the flow of modern mathematics. It is one thing to suggest an approach, but without motivation it may be hard to begin the work to make the dream come true. It is easier to prove a theorem when you are convinced of its truth, and for many modern mathematicians Odlyzko's computations have provided that important psychological foundation.

The work of Montgomery, Dyson, and Odlyzko focuses attention on the spacings rather than the levels, thereby reversing figure and ground. From the point of view of spacings, zeta zeros seem to behave exactly like the eigenvalues of your average random matrix. So, now we return to Hilbert and Pólya's earlier idea: Could it be that there is a specific matrix out there with the property that not only are the spacings between its eigenvalues like those between the zeta zeros, but, even more significantly, that among the morass of matrices there is one whose eigenvalues could be the zeros of Riemann's zeta function?

Pólya is famous for saying in *How to Solve It:*

> *If you cannot solve a problem then there is an easier problem that you can solve. Find it.*

Perhaps the search for that fundamental matrix whose eigenvalues are the zeta zeros is this simpler problem. The hunt was on.

12 *God Created the Natural Numbers . . . but, in a Billiard Hall?*

WITH THE GOAL of settling the Riemann hypothesis, many a mathematician set off in search of a path connecting the probabilistic with the particulate, a way to bridge quantum mechanics and the primes. Odlyzko's experiments seem to supply undeniable evidence that at the heart of the Riemann hypothesis beat the vagaries of chance, wrapped caduceus-like around the constancy of number. But the Dyson-Montgomery-Odlyzko law sets up a sort of Riemannian roulette: the eigenvalues of a randomly chosen Hamiltonian matrix appear to have precisely the same statistics as the zeta zeros. Where do you place your bet if you're hoping to hit prime number pay dirt?

Well, as it happens, at the same time as uncertainty and certainty were meeting in the Riemann hypothesis, they were being linked in a whole new way within the world of physics. Here too the path would run through the tangle of ideas that embody random matrices. Taken together this web of ideas would suggest to many a new physics-inspired trail to a resolution of the Riemann hypothesis, worthy of pursuit.

This route to Riemann is one that follows the axis of scale, moving between the macro and the micro, thereby bridging the phenomenological divide that seems to exist between determinism and chance that is characteristic of the worlds of classical and quantum physics. So we now arrive at a nexus for mathematics and physics as number and geometry are knit with quantum mechanics and chaos theory. This is a land at the cutting edge of research on the Riemann hypothesis, a place in the landscape of science where the Holy Grail—a solution—

may lie. It is only fitting that at this point in our story we are led by Sir Michael Berry, a man who is both a Knight Bachelor of the British Empire and a physicist, whose special mixture of brilliance, courage, and dash is helping him to lead a charge at the boundaries of science. This is the realm of ***quantum chaos,*** and it can be found at the end of a road that is the ***semiclassical limit.***

Life at the Semiclassical Limit

As we navigate the path from smallest of small to largest of large, the laws of physics move in tandem, morphing from the mosh pit of fundamental particles inside the atom to its celestial counterpart that is the dance of heavenly bodies in our solar system. In this way physics executes a journey which embodies the famous *correspondence principle* of Niels Bohr—that for every quantum system there is a corresponding classical system and vice versa.

Part and parcel of any quantum mechanical system is *Planck's constant,* named for the German physicist Max Planck (1858–1947), who was responsible for so many of the experiments that helped carve out quantum theory. Like the speed of light, Planck's constant* is one of the fundamental numbers of the physical world. It quantifies the vicissitudes of a quantum system, that aspect of the mysteries of subatomic behavior embodied in Heisenberg's famous "uncertainty principle."

Heisenberg's uncertainty principle is usually summarized as the impossibility of the simultaneous exact measurement of the velocity and position of a particle subject to quantum mechanical rule. More precisely, the uncertainty principle quantifies the extent to which both can be measured, giving a number that describes the least possible amount of error that could be incurred in taking these simultaneous measurements, regardless of how careful the measurer may be. This number is Planck's constant.

Were it zero, then both could be measured perfectly, but since it is greater than zero, it serves as a reminder of the limits of our knowledge on the small scale, a consequence of the fact that the mechanisms

*Planck's constant is equal to 6.6262×10^{-34} joule-seconds.

needed to measure phenomena in this domain must necessarily disturb something as they measure it.

Classical mechanics has no such uncertainty, so in order to study the boundary between classical and quantum mechanics we reconsider the laws of physics under the assumption of an ever-shrinking value for Planck's constant. This is life at the semiclassical limit, and it is akin to a gradual relaxation of the constraints on our knowing, which in the limit, as this number gets ever closer and closer to zero, entails no bound at all on the precision of our knowledge. The situation that we now hypothesize is effectively replicated as we measure quantum phenomena whose magnitudes dwarf Planck's constant, thereby making its effects negligible. In such a setting, the laws of quantum mechanics dissolve into those of classical mechanics, like the phase transition that occurs as steam turns to water, or water to ice, or vice versa. Moreover, this is the situation in which we find ourselves as we measure the spacings between energy levels of the nucleus at its outer reaches, the same measurements that helped Dyson, Montgomery, and Odlyzko bring quantum mechanics to the primes. The road to the Riemann hypothesis has now led us to this border territory. It's a strange place, but it's a place where Sir Michael Berry (b. 1941) is at home.

A KNIGHT OF THE BILLIARD TABLE

Berry has said, "At the risk of sounding slightly paradoxical, I would say that we are discovering the connections between classical mechanics and quantum mechanics to be richer and more subtle than either mechanics is when considered on its own." Indeed, borderlands often are places where much action occurs. These can be complicated regions seemingly outside the laws of any defining neighbor state, but, nevertheless, they still have their own internal logic.

If we are to believe the movies and history books, the leaders in these places that are neither one nor the other are often a strange hybrid of visionary and pragmatist: tough enough to endure hardships, but romantic enough to see possibility where others see complete confusion; confident enough to believe in their own success, but not so pigheaded as to ignore or disdain help wherever it might be found. These were and are people for whom contradiction does not exist or, at the very least, does not matter. These same qualities are necessary for success at a scientific boundary too, so it is not surprising that they can be found in Berry, currently Royal Society Research Professor at Bristol University and a leading researcher at the quantum-classical boundary.

Berry appears to be an incredible blend of opposites. He is of a hard-scrabble and decidedly nonacademic background, but was knighted for his contributions to fundamental physics. As a speaker he is pursued as much for after-dinner speeches as technical lectures. He has been both industry scientist and university professor. His writings reveal a man as comfortable reviewing a play or book as explicating a difficult piece of mathematics, so that of the hundreds of papers of which he is the author or a coauthor, nearly as many concern the arts as the sciences.

This is the mixture that Berry brought to bear on a problem which in the 1970s had begun to appear as one of the mysteries of modern physics. How does chaos manifest itself in the quantum world? What is quantum chaos? In looking for the answer to this question, Berry and the world of physics would find the Riemann hypothesis.

Quantum chaos

It is one of nature's outstanding ironies that in spite of the indeterminacy seemingly built into the probabilistic world of quantum mechanics, it is not by and large a place where we find ***chaos***—at least, not chaos as we now understand it in our own more familiar everyday experience.

The archetype of chaos is the "flapping of the butterfly's wings," a hypothetical, poetic perturbation in the airflow over Brazil which might initiate, in the following days, a catastrophic tornado in Texas. The way a tiny disturbance might engender dramatic changes in the evolution of the weather is now the popular paradigm of what mathematicians and physicists would call ***sensitive dependence on initial conditions,*** a hallmark of mathematical chaos and ***chaotic dynamical systems.*** Clothed in metaphor, it is a theme explored time and again in literature and film, one which resonates with each of us as we ponder roads not taken, choices made and not made, and the way in which seemingly insignificant events can have unforeseen and surprising ramifications. Fumbling for change at a newspaper stand may have caused you to meet the love of your life, or instead to miss a train that subsequently derailed.

Sensitive dependence is but one characteristic of chaotic systems. Another is ***ergodicity,*** a word used to describe the property of a system for which almost all initial conditions mark the origin of paths that over time eventually explore almost every region of space. This is the aspect of the chaos of life reflected in the timeworn adage, "If you live long enough, you'll see just about everything."

While chaos may indeed rule the course of human events, it did for some time appear to be irrelevant to nuclear events. At least this was the thinking in the scientific world in the early 1970s. The nanosecond timescale of the quantum world would appear to rule out phenomena characterized by long-term behavior; the smooth wave functions that describe quantum mechanical situations exhibit the same discrete set of energy levels no matter what sorts of external disturbances they experience. Theirs is a virtual steady state of change, seemingly immune to perturbation.

Nevertheless, according to Bohr's correspondence principle, every classical system has a quantum analogue, so that when the principle is applied to a chaotic classical system, somewhere in its quantum reflection must twinkle the specter of disorder. Somewhere in the most energetic reaches of the quantum world, where Planck's constant pales in comparison with the measurements of phenomena which it is bound to constrain, the strange order inherent in quantum mechanical systems must dissolve into chaos. This is that part of the borderland where chaos meets quantum mechanics, where we find the real mystery, and possibly where the key to the Riemann hypothesis may lie. This is the land of quantum chaos.

BILLIARD TABLES FOR PHYSICISTS

In many a border town, the local pool hall might serve as the place where folks go to unwind, relaxing after a long day on the job, soothed by the gentle clicks and clacks of ball meeting ball, or the quiet thump of ball hitting bumper. Such a place is the Chaotic Cue, tucked away on a small side street in our mythical village of Quantum Chaos. It is the local hangout for the roguish physicists and raffish mathematicians who populate this dusty frontier town. It is a place where many are taking and placing bets that the Riemann hypothesis will play itself out on the surface of a billiard table.

As you enter the establishment, you discover toward the back of the room an archipelago of billiard tables whose shapes vary from familiar rectangles to bizarre fractals. The most basic laws of physics, classical mechanics known even by Euclid, guide the trajectories of the billiard balls constrained to move within these felt-filled islands in the front room. Built by the most discerning and careful of physicists, the frictionless table surfaces enable the perfectly struck smooth spheres to careen about the tables for as long as we allow them. In the lingo of physics they behave as free particles subject to no external forces, whose angles of reflection upon striking a bumper equal their angles of incidence. It is a spectacle of *specular reflection.*

We survey the room and wander over to watch the younger set who seem to be sticking to familiar rectangular, elliptical, and circular tables. We pause at a rectangular table and watch as a first-year graduate student chalks her cue and strikes the ball. It hits the far bumper at an angle of forty-five degrees and bounces off at the same angle. After ten bounces against the bumpers it arrives back where it started, having traced a *closed trajectory.* We do not stop its motion and it continues on and on, executing the same ***periodic orbit*** over and over again. An example is shown in Figure 33.

We move on to the circular table across the way. Here a second-year sharpie slathers chalk on his cue. He aims, pulls back his cue stick, and strikes the black billiard ball with all his might. The ball goes zipping about the table, leaving behind a jet stream of chalk that traces its trajectory. We watch in awe, mesmerized by the beautiful Spirograph-like pattern that gradually appears, a trail left by a billiard ball dutifully obeying the laws of classical physics. For all the intricacies of the design left in the chalk-filled wake of the ball, there is one constant: the ball stays free of the center of the table, leaving untouched a region outlined by a smaller concentric circle. Indeed, upon close observation we see that in fact as the billiard ball makes its way across

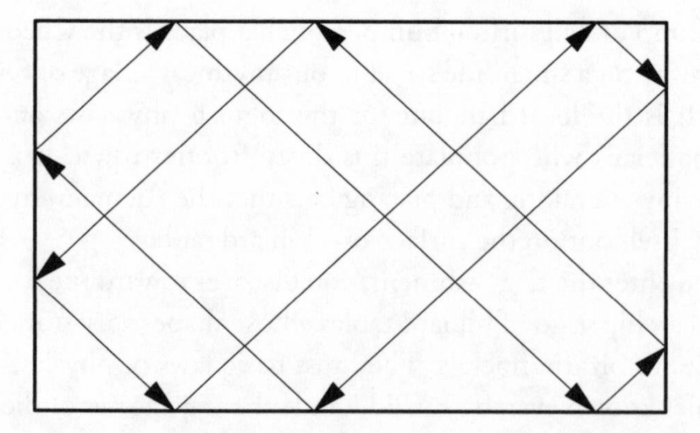

Figure 33. Following the path which starts anywhere on this trajectory will bring you back to the beginning after ten bounces—it is a perfect closed trajectory. (Figure from P. Kostelec.)

the table, the arc that it traces out just brushes alongside this central no-ball's land, lying tangent to the inner circle's boundary. This phenomenon is illustrated in Figure 34.

Delighted with the beautiful patterns in the circle, we turn to a nearby elliptical table where the graduate students are working the angles. The behavior here is much like that on the circular table. In

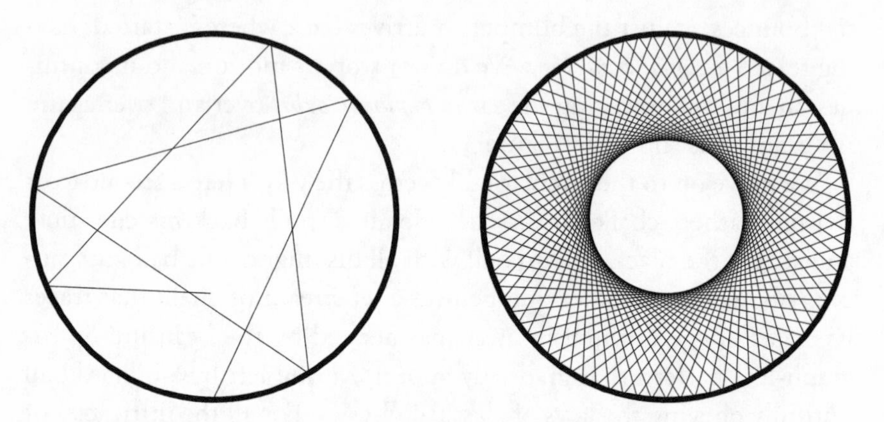

Figure 34. At left are the first few bounces of the billiard ball on the Euclidean circular table, and at right is the pattern that appears after many, many more caroms. Notice that a central, concentric region remains untouched, although each time the ball goes zipping across the table it just brushes this region, tracing a line tangent to the inner circle. (Figure from P. Kostelec.)

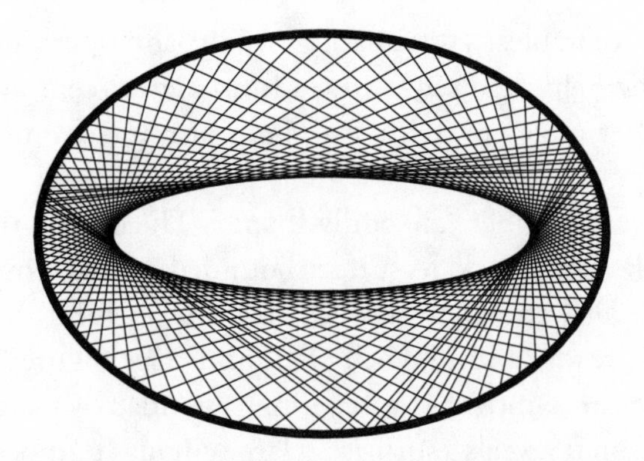

Figure 35. Path of a billiard ball bouncing about an ellipse. The ignored central region is defined by a confocal ellipse; "confocal" means that the interior ellipse has the same foci as the exterior ellipse. This is the elliptical counterpart of concentric circles. (Figure from P. Kostelec.)

direct analogy to the circular table, there is a deserted central region defined by a *confocal ellipse.* Figure 35 shows this.

These beginners are all playing on tables where the flat surfaces and linear boundaries defining the geometry of the table create laws of motion of an ***integrable system,*** a name that comes from the ability to predict the trajectories of motion through the calculus-based technique of integration. In these cases the untouched central regions are evidence of a certain *constant of motion* that constrains the behavior of the bouncing balls. This is a condition that rules out chaos, ensuring that two balls which differ only slightly in their initial conditions—a difference, say, that was caused by a butterfly hovering near the table—will still stay fairly close together, tracking each other like two dogs running playfully side by side down a path in the woods. These are the simplest of dynamical systems, paradigms of regular and predictable behavior. Chaos, however, can be found just around the bend.

Russian Billiards

We continue to walk around the room, and eventually we come to a pair of less familiar-looking billiard tables. In contrast to the torpor surrounding the predictable motions on the integrable tables, here we find a much more animated and disorderly group. Scientists circulate about a pair of tables, gesticulating wildly and speaking loudly in Russian. They help themselves to tea from a great central samovar as well as a buffet table covered with plates of blintzes and caviar, and carafes of pepper vodka.

The billiard tables here are oddly shaped. The one on the left looks something like a racetrack, its surface bounded by two semicircles connected by parallel straightaways. On the wall behind this table is a plaque which reads "Bunimovich's Stadium" (see Figure 36). On the other side of the samovar is a seemingly standard square table, but upon inspection it reveals a surprise: a large circular bumper in the center. The plaque above this table reads "Sinai's Table" (see Figure 37).

The trajectories of the billiard balls on these tables exhibit much more variety than those executed on the integrable surfaces. While simple periodic motions are possible (e.g., the simple back and forth of a ball in the stadium-shaped table which ping-pongs forever between the parallel sides), the general path is much less predictable. Many a ball executes ergodic motions, tracing out paths which over a long time would explore almost every region of the table.

The play on these tables is more like the weather over Bermuda. The course of these billiard balls is also subject to the flapping of the butterfly's wings: were butterflies let into the Chaotic Cue, the tiny air eddies they would initiate near these tables could cause enough varia-

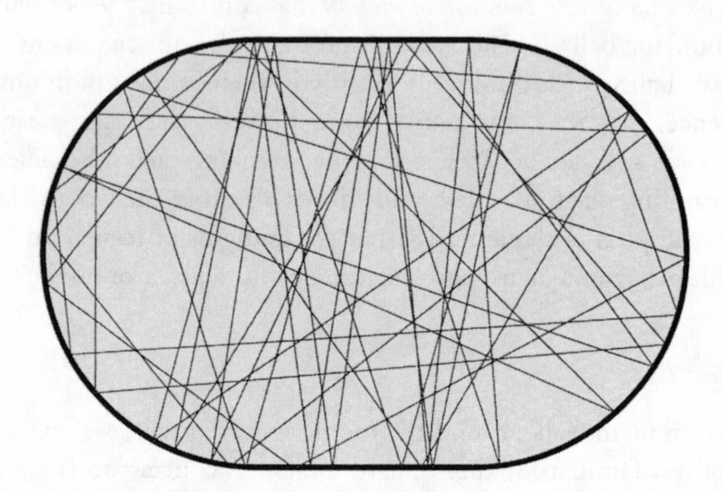

Figure 36. Bunimovich's stadium, with the beginnings of one of the general ergodic paths. Chaos ensues. (Figure from Eric Heller.)

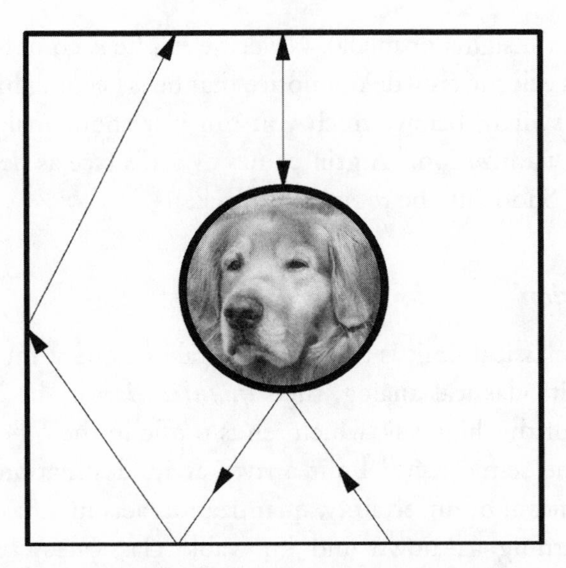

Figure 37. Sinai's table is a square boundary with a circular bumper in the middle, in this case embossed with a picture of Digger, the mascot of the Chaotic Cue. While periodic trajectories are possible, for example the back and forth indicated by the two-headed arrow on top, the general trajectory is one that fills up the open region between the boundary and the circle; it is an ergodic path. (Figure from P. Kostelec.)

tion in the starting points of two different trajectories that after some time these trajectories would differ dramatically.

Sensitive dependence and ergodic motion are two of the defining characteristics of a chaotic ***dynamical system.*** Sinai's table (named for the Russian mathematical physicist Yakov Sinai, now a professor of mathematics at Princeton University) and Bunimovich's stadium (named for Sinai's former student Leonid Bunimovich, now a professor at the Georgia Institute of Technology) are two classic examples. But neither of these new strange examples has prepared you for what will come next.

You take a step back, sip your vodka, relax, and take in the scene. Off in the corner of the hall, through the smoky haze, you see a portrait. You approach it, and sure enough, there he is: Riemann. Below, a small sign reads "This Way to Riemann." A few paces away is a kindly-looking, balding, bearded gentleman perched on a stool. He seems to be overseeing the action at the tables. He appears overdressed for the roughneck room, and indeed appears to be wearing a sash with some sort of insignia or medal. Nevertheless, he is completely at ease among this eclectic crowd. You notice that he is keeping his hand on a small wall switch, below which you can just about make out some writing: "Quantization." A grin comes over his face as he reaches for the switch. Suddenly the room goes black.

Quantization

If the semiclassical limit is the road that takes a quantum mechanical system to its classical analog, then ***quantization*** is the lane on the other side of this highway, which sends traffic in the other direction. Whereas the semiclassical limit arrives at its destination through a gradual removal of uncertainty, quantization acts in a flash, instantaneously turning a known and knowable classical system into an uncertain quantum system.

Quantization is generally a complicated mathematical procedure that is more art than algorithm; but in the context of our billiard table it takes a well-documented form, trading the well-defined certain nature of the moving particle for the uncertain form of a quavering wave. This has the effect of transforming our games of billiards—in which we want to predict the trajectories that will fit on a given table—into a nanoscale analogue in which, instead, the goal is to determine the sorts of wave patterns that will fit perfectly into the various *planar domains* defined by the tables.

In this way the quantization of any of these billiard tables can be thought of as a process in which the billiard table is turned into a drum of the same shape. Instead of striking a ball and watching it carom about the table, we bounce a billiard ball off the drumhead and listen to the sound caused by the ensuing wave. The perfect waves are the *standing waves* that fit the drumhead. In Figure 38 two such examples for circular domains are shown. Were this problem studied in one dimension, the planar drumhead would be a replaced by a single guitar string, and the standing waves are precisely those finite portions of sine waves that fit exactly, say from one crest to another, between the string's two end points. The number of complete waves (counted from trough to trough, or equivalently from crest to crest) that a single standing wave fits in the string is its frequency. In an analogous

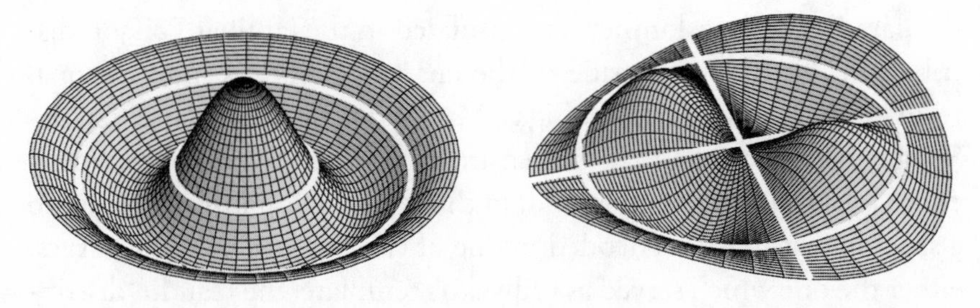

Figure 38. Two different standing waves in a Euclidean circular domain. If animated, these waves would simply move up and down, as opposed to sloshing from side to side. (Figure from P. Kostelec.)

way our two-dimensional standing waves fit the drumhead. They also have a frequency; and the higher the frequency, the more energetic is the wave.

The different shapes of the domains defined by these billiard tables imply that different kinds of standing waves fill them. But for all the different sorts of waves that can appear, some general patterns begin to emerge, patterns that appear in the distribution of the spacings of the energies with which the drumheads resonate, just like the spacings that Odlyzko used to connect the Riemann hypothesis to randomness, matrices, and nuclear energy levels.

Maybe this is why, as the lights suddenly come back on and the din of the submicroscopic tom-toms is replaced once again by the muffled collisions of ball with bumper and the click-clack of ball against ball, that man by the light switch, Sir Michael, is grinning.

The basic conjecture of quantum chaos

In work with Michael Tabor (professor of applied mathematics and physics at the University of Arizona), Berry formulated the first half of what has come to be known as the ***basic conjecture of quantum chaos:*** the spacing distribution for the energies of the standing waves in a drumhead are determined by the dynamics of the corresponding billiard table and, in particular, are of one of two kinds, depending on whether the dynamics are integrable or chaotic.

Berry and Tabor conjectured that integrable classical systems would have a quantization in which the spacing distribution follows the paradigm of randomness encapsulated in the familiar Poisson distribution. On the other side of the chaotic divide, an international team of physicists—Oriol Bohigas, Marie-Joya Giannoni, and Charles Schmit—conjectured that the sensitivity to initial conditions experienced by a chaotic system would, in the semiclassical limit, give rise to a spacing distribution encoded in one of two smooth bell-like curves: either the one which served as Odlyzko's template, the Gaudin distribution, which comes from the GUE distribution, or a cousin that comes from the GOE (Gaussian orthogonal ensemble) distribution.

In this conjecture we see a sort of turning of the tables taking place at the semiclassical limit, in which the chaotic and the integrable seem to trade places. Here chaos manifests itself in the rigidity implicit in the well-defined gaps between the energies. On the other hand, integrable systems produce spacing distributions without rule. Once again we can derive some intuition for this topsy-turvy physics in the laboratory of our own experience. Think of real-life systems that are, or are not, dramatically affected by small changes. For instance, the "sensitive dependence" characteristic of your most rigid friend, for whom even the smallest deviation from routine is enough to engender tremendous tumult, versus the ease with which a laid-back buddy takes in stride all the changes and randomness that life provides.

A key tool used in formulating the basic conjecture is another amazing link between the classical and quantum sides called a ***trace formula.*** First proposed by the physicist Martin Gutzwiller of IBM in the 1960s, this trace formula gives a direct mechanism for relating the lengths of certain "prime" (in the sense of primal or fundamental) periodic billiard trajectories to the resonant energies of their wavy quantized cousins. Typical examples are phenomena such as the simple back-and-forth trajectories possible on Sinai's chaotic bumper pool table. Gutzwiller's formula shows that the numbers obtained in summing lists of the lengths of these trajectories are approximately equal to analogous sums of certain groups of frequencies, the latter of which are called ***traces.***

The basic conjecture of quantum chaos is a huge first step toward distinguishing among the Hamiltonian matrices that Pólya and Hilbert first suggested might lead to a confirmation of the Riemann hypothesis. It is a complicated chain of reasoning, but one which links a lot of research. Remember that Pólya and Hilbert had been the first to suggest that the zeta zeros might be regarded as the eigenvalues of Hamiltonians that had a very general Hermitian symmetry, which are in turn the energy levels of a quantum system described by the Hamiltonian. Odlyzko then indicated that the distribution of the spacings of the zeta zeros are well matched by the smooth curve of Gaudin's GUE distribution. But assuming the basic conjecture of quantum chaos, energy levels whose spacing distribution has this sort of shape

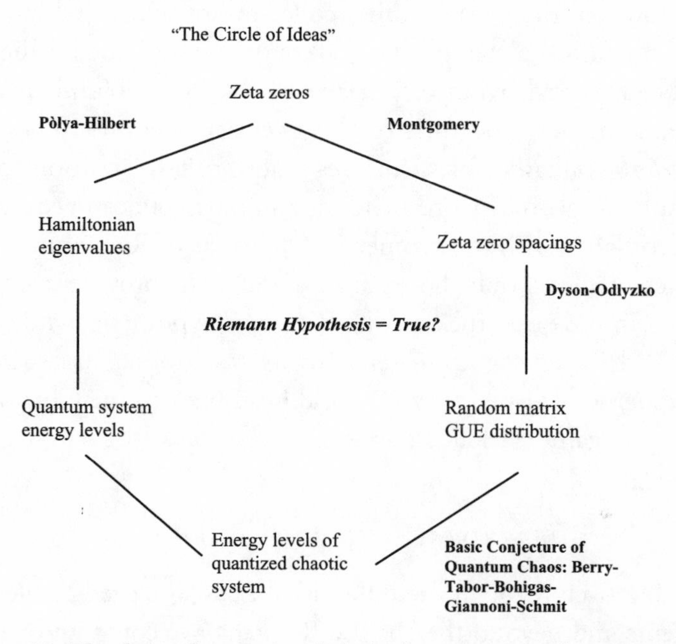

Figure 39. The chain of connections linking the zeta zeros and quantum chaos. Polya and Hilbert first dreamed up the idea that there might be a matrix whose eigenvalues are the zeta zeros. If this matrix is the Hamiltonian of a quantum mechanical system, these eigenvalues have the interpretation as energy levels. Evidence comes from Odlyzko's calculation of the the spacing distribution for the known zeta zeros. It shows extraordinarily good agreement with another distribution of numerical differences: those derived from the eigenvalues of a randomly chosen Hermitian matrix, also called "the GUE distribution." This motivates a search among the Hermitian matrices for a Riemann Hypothesis-settling matrix. The Hermitian matrices are conjectured to be precisely those that are characteristic of the Hamiltonians that come from quantum versions of classical systems that are "chaotic." This claim is one half of the "Basic Conjecture of Quantum Chaos" and thus motivates a search for a Riemann Hypothesis-settling quantum chaotic system.

would come from a quantized chaotic system, encoded as the eigenvalues of an associated Hamiltonian matrix. Thus, if we're looking for zeta zeros masquerading as energy levels or eigenvalues of a Hamiltonian, we don't need to investigate all kinds of Hamiltonians, just those that describe quantized chaotic systems. Phew!

The first places to look were those settings in which quantum chaos originally appeared. Alas, using tools like Gutzwiller's trace formulas, it was easy to show that neither Sinai's table nor Bunimovich's stadium would give energy levels appropriate to the Riemann hypothesis. However, there was another collection of billiard tables that were also known to be chaotic. They were of an auspicious pedigree, having been crafted by an old master of the Prime Number Theorem, Hadamard, near the turn of the twentieth century. These are the ***hyperbolic*** billiard tables, surfaces whose rules of motion follow not the familiar geometry of Euclid, but a non-Euclidean geometry more in accordance with the geometric inventions of Riemann. The games at these tables would be exciting enough to provide another cliffhanger in the quest to settle the Riemann hypothesis, while inspiring new and intriguing approaches to its resolution. So, we'll pick up what's left of our pepper vodka and head back to the Chaotic Cue to watch the games at Hadamard's tables.

BILLIARDS IN THE POINCARÉ DISK

Walking back through the billiard hall, past the predictable graduate students and beyond the chaotic Russians, we come upon a fashionable set of Europeans. In this part of the room men and women are laughing, talking, drinking red wine and espresso, and making periodic trips to a buffet table filled with bread and a selection of cheeses. A few are smoking Galois cigarettes. Looking quite forlorn and very out of place is what appears to be a time traveler, a white-haired man in sandals and a robe. Who could he be?

This part of the game room is filled with tables that are nearly triangular, whimsically shaped islands of felt, bounded by three gently curved sides of varying lengths and curvatures. Amid these graceful green divots is a large circular table. It seems to be the site of some sort of demonstration, and we walk to it.

We peer over the hushed crowd to see a player placing three billiard

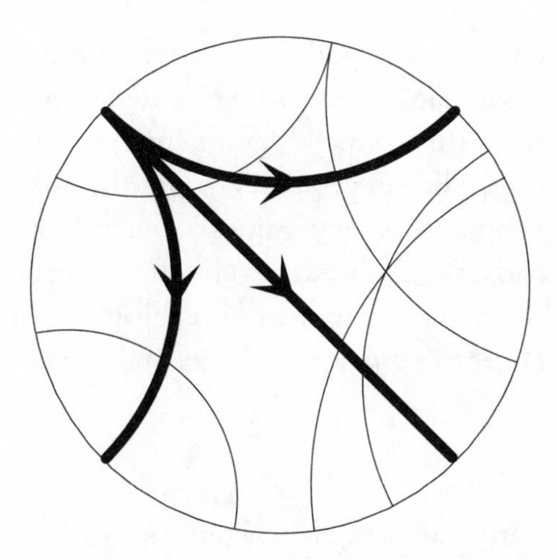

Figure 40. In bold outline are the full trajectories of the three distinct billiard balls sent spinning along the strange non-Euclidean table. The central straight line marks a diameter of the disk. The sweeping arcs on either side are but pieces of circles that would extend beyond the boundaries of the table. In particular, notice that these three very nearby starting points of balls pointed in nearly the same direction give very different trajectories—a hallmark of chaos. The other faint curves show the trajectories traced out by other imagined balls sent on their way from the edge of the table. (Figure from P. Kostelec.)*

balls almost side by side at the edge of the table. She chalks the cue and first taps the middle ball, sending it straight ahead on a diameter to the other side, where it bounces right back to her. She then addresses the ball on the right. We watch closely, fully expecting the ball to trace a chord running alongside the first diameter, but no! Instead, the ball veers off to the right, drawing a beautiful circular arc that falls away from the first straight path. This ball hits the side at a right angle, and rebounds back directly along the same arc, coming to a stop at its starting point! She moves to the remaining ball on the far left, and a similar stroke creates an arc which is the mirror image of the last shot, falling away from the diameter, caroming off the side, to once again return to its starting point. These three trajectories are illustrated in Figure 40.

Our billiard master places her cue stick on the table and makes a

*To be precise, the geometry is such that each ball would take infinitely long to reach the end of its path, so that it approaches the bumper in the limit, as does the person who crosses a room by successively halving the distance between him and the opposite wall.

long sweeping bow. At this point the crowd erupts in applause, all except for our sad sandaled friend, who now looks even more unhappy than before. Indeed, this is Euclid, and here in front of him is proof that his geometry, Euclidean geometry, is not the only game in town. All around him people are playing games of billiards on tables cut from this cloth that bends straight lines into arcs. It is evidence of ***hyperbolic geometry,*** and he has just witnessed an exhibition on the decidedly non-Euclidean ***Poincaré disk,*** where, in general, chaos rules the day.

The Poincaré Disk

Named for the mathematician Henri Poincaré (1854–1912), the Poincaré disk is a self-consistent geometric world which provides a model for a non-Euclidean geometry, a mathematical world of points and lines in which our usual Euclidean notion of parallelism does not hold.

Poincaré is considered by many to be one of the most creative scientists of all time. He was a codiscoverer (along with Einstein and Lorentz) of the theory of special relativity and the father of the fields of dynamical systems (an outgrowth of his work on the famous three-body problem) as well as modern topology. In particular, he is the author of a topological cousin of the Riemann hypothesis, the *Poincaré conjecture,* which characterizes those sorts of objects that can be continuously transformed (i.e., "morphed") into a three-dimensional sphere.* Like the Riemann hypothesis, the Poincaré conjecture has a $1 million bounty, and the search for the resolution of this topological mystery has led to several Fields Medals and a huge body of elegant and beautiful mathematics.

But as regards the quest to settle the Riemann hypothesis, all that

*Here, "dimension" refers to how many numbers we need to reference a position on the object of interest. So a "one-dimensional sphere" is a circle, since we need only one number—given by the angle measured off a horizontal diameter—to indicate a position on a circle. A "two-dimensional sphere" is like a ball, as we can use longitude and latitude to describe a point. Furthermore, notice that a one-dimensional sphere (circle) can be carved out of two-dimensional space, described as all the points at a fixed distance away from a center. Similarly, a two-dimensional sphere is carved out of three-dimensional space, again consisting of all the points at a fixed distance from a center. So a three-dimensional sphere is in four-dimensional space, made up of all points a fixed distance from a center. Yes, it is very hard to visualize.

we care about is the Poincaré disk. Like a world observed through a microscope, it is a land contained entirely within the interior of a circle; and like the microbial worlds we often find on the other end of the microscope lens, it is a beautiful but strange and surprising world.

Play on the Poincaré disk shares with each of the billiard tables in the backroom of the Chaotic Cue the basic principle that the balls follow ***geodesics,*** paths which between any two points give the shortest distance (assuming that there was no carom in between). In the familiar Euclidean world geodesics are the Euclidean straight lines, but as evidenced by our pool shark, there must indeed be worlds in which straight appears anything but that. So when our pool shark was performing her demonstration she was certainly a straight-shooter, but shooting straight in a non-Euclidean way.

The non-Euclidean/Euclidean divide traces its origins to Euclid's systematic and objectively intentioned investigation of geometry, *Elements.* There he sets down five seemingly self-evident axioms regarding the fundamental geometric objects that are the point and the line. The first four axioms are little more than definitions of point and line and the relations between them. The fifth axiom, however, is an outlier, a declaration that through any point off a given line, exactly one line parallel to the original can be drawn. For centuries mathematicians tried to show that the first four axioms imply the truth of the fifth—that in effect, this property of parallelism depended on the original four axioms; and hence, that in order to have a logically consistent geometry it could not be otherwise. The failure of this quest, indeed a proof of the exact opposite, that the parallel postulate is independent of the original four axioms of geometry, is one of the most significant mathematical achievements of all time. It was a result so remarkable and controversial that when Gauss first discovered it, he kept it secret and did not announce his result until after two other mathematicians, Bolyai and Lobatchevsky, had broken the ice with their own claims of priority.

So non-Euclidean geometries take the parallel postulate of Euclid and ask, "What if it were otherwise?" That is, can we come up with logically consistent systems of entities that we call points and straight lines, in which all things that Euclid posits hold, except that we permit the parallel postulate to be different?

An equivalent formulation of the parallel postulate is the declaration that the sum of the angles of all triangles must be 180 degrees. A non-Euclidean geometry will have triangles whose angles sum to either something less or something more than this. It is a wonderful irony that the true geometry of our own earth provides an archetype for non-Euclidean geometry. Gauss discovered this, most likely as part of his work as a land surveyor, noting that a huge triangle on the earth—say one whose vertices are the cities of New York, Paris, and Caracas—will have angles whose sum is greater than 180 degrees, a consequence of the curved surface of our planet.

As it should be, this global triangle is demarcated by the intersections of three distinct "straight" lines. But here it must be the case that "straight line" means something different from what Euclid intended, for a Euclidean straight line between any two of these cities would necessarily cut through the earth. Of course what we demand here is that these shortest paths between cities lie on the surface of the earth. If we think of these paths as following along a string laid between the cities, the shortest path would be determined by that string, which has no slack. Using a perfect sphere to approximate the earth, you may be able to convince yourself that the shortest paths joining the vertices of our international triangle are arcs contained within the ***great circles*** on our earthly sphere, rotated versions of the equator which are determined by the choice of two (nonantipodal) points (say New York and Paris). In fact, such considerations account for current transoceanic flight plans where "shortest path" means "cheapest path."

In this setting Euclid's fifth axiom fails miserably.* Any two different great circles intersect at exactly two points, so that given a particular great circle, and a point not on it, it is impossible to find a great circle through the orphaned point that does not intersect the original one. This is a decidedly non-Euclidean geometry.† More precisely, it is called an elliptic geometry, derived from the Greek word *elleipein,*

*Not only does the parallel postulate fail, but so does Euclid's second axiom, which declares that lines must be infinite in extent.

†The precise definition of this spherical geometry requires a slightly more complicated construction. A *point* is defined as a pair of antipodal points, like the North and South poles. A *line* is an arc of a great circle, and since any two great circles either are coincident or intersect at exactly two antipodal points, we do indeed find that every two lines intersect, so there are no parallel lines.

which means "to fall short," a phrase that we might use to describe the shrinking distance between two ships starting at the equator and simultaneously heading to the North Pole along these spherical geodesics.

In opposition to the sphere is our Poincaré disk. Here straight lines come in two flavors, both nicely demonstrated by our friendly pool shark. On the one hand, diameters of the disk are permitted, so if the two points lie on a diameter of the disk, then this is the geodesic connecting them. This corresponds to the pool shark's first shot, straight across the table and back again. Our shark's second two shots show the new and interesting geodesics. This time as she sent the balls "straight" off the boundary, they traced out the beginnings of beautiful half-moons, first to the left and then to the right. Indeed, if two points don't lie on a diameter, then it is possible to draw only one circle in such a way that both of these points lie on its circumference and it intersects the disk at right angles. In this sense two points on a diameter are viewed as also lying on a (second) circle of infinite radius. The arc connecting the two points in the interior of the Poincaré disk is then the geodesic between them. A nondiameter geodesic is shown in Figure 41.

How these balls on either side of the diameter veer away from one

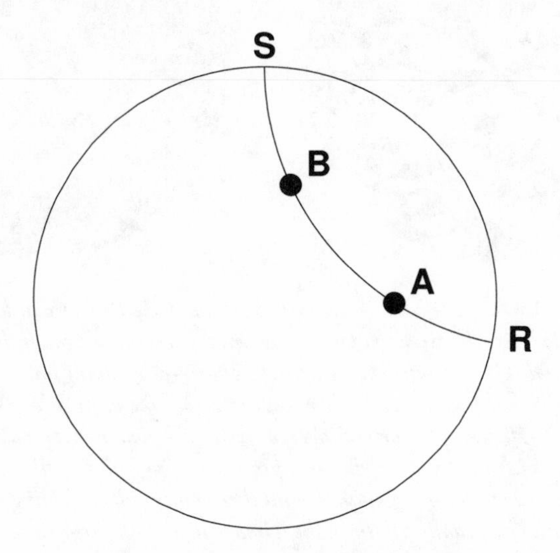

Figure 41. A surprising straight line in the Poincaré disk. It is the only arc from a circle that goes through these two points while intersecting the boundary circle at right angles. (Figure from P. Kostelec.)

another or fall beyond is summarized in the Greek word *hyperballein,* meaning to "throw beyond," providing the source of the name ***hyperbolic geometry*** to describe the interaction among points and lines in this curvy world.

Thus in the Poincaré disk, given a line, say the horizontal diameter of the disk, and a point not on that line, an infinite number of hyperbolic straight lines can be drawn through that point in such a way that the two lines never meet. In other words, an infinite number of lines parallel to the original diameter can be drawn through the outside point. This is the source of its designation as non-Euclidean.

It is an alluring world that has entranced many an artist, most notably M. C. Escher, whose wonderfully symmetric mosaic disks are filled to the rim with patterns that are the hyperbolic analogue of the simple repeating patterns of a Roman or Greek frieze. Here instead Escher takes an initial pattern and repeats it in the Poincaré disk by reflecting it straight (hyperbolically) across hyperbolic straight lines. Examples are shown in Figure 42.

Figure 42. On the left is a tiling or tesselation of the Poincaré disk by hyperbolic triangles (the triangles formed by three intersecting straight lines in the Poincaré disk). The entire picture can be obtained by taking one of the triangles at the center and repeatedly reflecting across one of its sides. Notice that this is a reflection performed over a geodesic—i.e., a hyperbolic straight line—at each border. Carrying this process out "to infinity" fills the disk, in the same way that taking a square and then continually reflecting it about its sides would eventually fill the plane. (Image used with permission of Jos Ley.) On the right (constructed by and appearing with permission of Silvio Levy) is an Escher-like tiling (seemingly inspired by Escher's Circle Limit, I*).*

The Cartesian plane, a sphere, the Poincaré disk—each provides a consistent model for a geometry, the last two in which parallelism, as we know it, fails. More intrinsically, these geometries differ in terms of their *curvatures.* The Cartesian plane is a *flat* world, a mathematical environment of zero curvature. A perfect sphere is a positively curved world, indeed one of constant curvature, as if each point were the peak of a perfect hill. This geometry has as its alter ego something like the Poincaré disk or *pseudosphere,* which is of a constant but *negative* curvature. A portion of the pseudosphere, shown in Figure 43, can be visualized as the surface of a trumpet. In this model of hyperbolic geometry geodesics can be found as on the usual sphere, by finding the "slackless path" traced out with string.

The paradigm of negative curvature is the *saddle point,* the place in the center of a saddle where in the direction from tail to head it curves up, and from side to side it turns down, while exploring all other curvatures in between (varying from minus one to one) as you move around the saddle. The constant negative curvature of the Poincaré disk is thus a world in which every point is a saddle. The Poincaré disk may seem more a dream than reality, but its three-dimensional analogue is one of the competing models for the large-scale structure of our universe.

Chaos in the Disk

In the hyperbolic geometry of the Poincaré disk nearby paths pointing almost in the same direction run away from one another at an exponential rate, their separation effectively doubling at each instant.

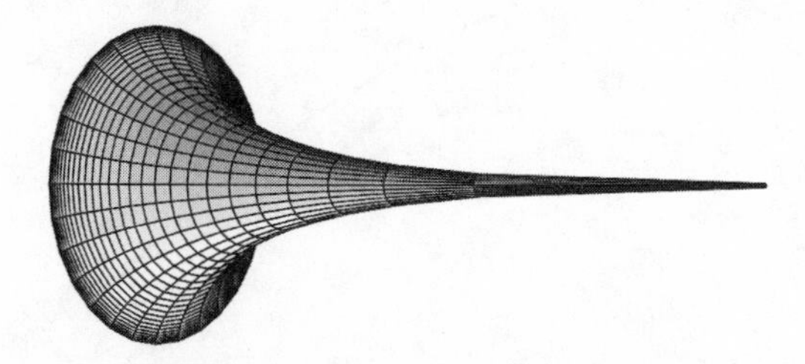

Figure 43. A portion of the pseudosphere. Here, given a line and a point off the line there are an infinite collection of lines through the remote point that do not intersect the given line. (Figure from P. Kostelec.)

This is easily visualized by imagining the different paths taken by two marbles set rolling from nearby locations on the trumpet surface. The curvature is such that the effect is less that of the butterfly and more that of the gigantic Mothra flapping her huge wings. This is the source of the sensitive dependence on initial conditions in this geometry, which was first noticed by Hadamard and which led to his discovery of the chaos that would ensue on the ***hyperbolic triangles.*** These are the divots which dot the French quarter of the billiard room, composed of three intersecting hyperbolic straight lines, arcs that create triangles whose angles sum to less than the 180 degree Euclidean version.

Hadamard determined that, as on our Russian tables, the general path traced out a hyperbolic triangle by this sequence of Riemannian-rebounds is an ergodic one, coming infinitesimally close to any point within the boundaries. Thus, this is an environment in which chaotic dynamics prevail. Figure 44 shows an example of such a triangle with the outline of the first few rebounds of such a triangle-filling trajectory.

Viewed afresh through the lens of the Berry-Tabor conjectures, the chaotic dynamics of these tables suggest that the waves which fit these tables (i.e., their quantizations), themselves subject to the constant

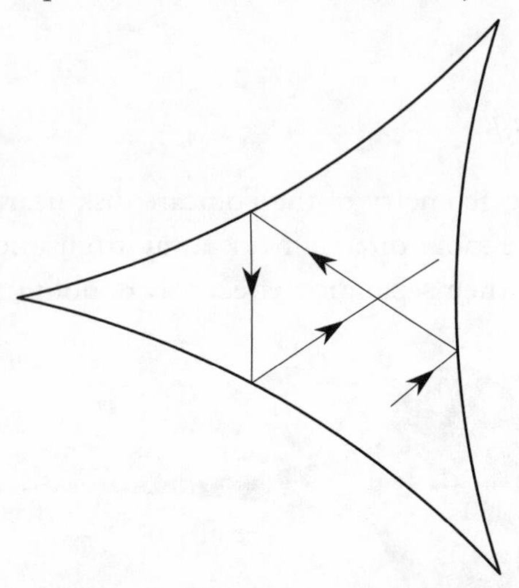

Figure 44. A typical equilateral triangle in the Poincaré disk. The arrows indicate the first three bounces of a free particle, i.e., one whose movements are simply tracing out the path dictated by the geometry of the region. (Figure from P. Kostelec.)

negative curvature of these islands, could have energies which are the zeros of the zeta function.

Selberg's trace formula

In this deep connection between geometry and prime numbers we find ourselves on a surprising intellectual geodesic. The first person to conceive of a mathematical notion of curvature was Gauss, who incorporated the idea into a consistent methodology for carrying out all the usual geometric measurements such as angle and length in non-Euclidean settings. These ideas were extended by Riemann in his invention of Riemannian geometry, and indeed were critical to the deep investigation of unfamiliar landscapes like the Poincaré disk. Could Riemann have known that he may have been creating the surveying equipment necessary for finding the solution to the Riemann hypothesis in the Poincaré disk?

By linking the Riemann hypothesis to dynamics in hyperbolic geometry, the physics community was rediscovering a connection that had first surfaced in the mathematics community nearly a generation earlier. In 1956 the Fields medalist Atle Selberg, the same man who had first "broken the positivity barrier" for the zeros of the zeta function (showing a positive proportion in the critical strip to be on the critical line), had already explored these hyperbolic hills and found a relation between the paths that crisscross them and the kinds of waves that wash over them. His discovery of what is now known as the *Selberg trace formula* could be one of the most important number-theoretic discoveries of the twentieth century, for among other things, it provides a possible template for resolving the Riemann hypothesis with the tools of hyperbolic geometry.

Whereas Riemann used eight pages to change the course of mathematics, Selberg takes forty pages in the *Journal of the Indian Mathematical Society* to present a surprising connection between geometry and analysis. In this paper he links the energy of the waves constrained to reverberate in the most general sorts of symmetrically defined regions of hyperbolic space and the periodic paths confined to a similarly shaped billiard table. The trace formula of Gutzwiller is but one of many, many mathematical children of Selberg's trace formula, each of which manages to find relatively simple relations between geometry and frequency.

At a remove of almost one century from Riemann's "On the Number of Primes . . . ," Selberg's paper is in many ways a natural descendant of this earlier mathematical masterpiece. Like its Riemannian forebear, it is an astounding amalgam of ideas which range over a diversity of mathematics. Upon a non-Euclidean and decidedly Riemannian landscape, it mixes the smooth wave phenomena of Fourier analysis with the hiccups suggested by certain "discontinuous" symmetries of the Poincaré disk—an echo of Riemann's own use of Fourier analysis in the investigation of the jagged and discontinuous accumulation of the primes.

One crucial tool in Selberg's work is the creation of his own zeta function, today known as the *Selberg zeta function,* which like Riemann's zeta function is a descendant of the harmonic series. Riemann took as his starting point Euler's rewriting of the harmonic series in terms of the primes. Selberg goes one step further, using the same idea but replacing the primes in the denominators by lengths of certain fundamental or *prime* trajectories by which billiard balls rebound in these hyperbolic regions. Whereas Euler first connected the distribution of the primes to lengths of strings laid out on a harmonious Pythagorean guitar, Selberg's modern geometric updating takes a non-Euclidean drumhead and connects the waves generated by banging this drum with the lengths of the "prime" trajectories that are the closed trajectories on the corresponding billiard tables.

Once again we might ask: What of the zeros of Selberg's zeta function? As in the case of Riemann's zeta function, it is simple to see that with only a few exceptions, the zeros must lie within a critical strip, which once again is split by a critical line. Here the truth of the analogue of the Riemann hypothesis is immediate—indeed, there is no distance between question and answer. Selberg goes on to show that for the Selberg zeta function all the nontrivial zeta zeros lie on the critical line.

But these are not just zeros of the Selberg zeta function; in fact, they are also naturally connected to the eigenvalues of a matrix. In particular, each of these zeros when squared and then increased by one-fourth becomes a measure of the energy in the standing waves able to live on these strange Riemannian surfaces. The Selberg zeta function has the whole package: zeta function, zeros on the critical line, and eigenvalues. For those following Berry's lead to the Riemann hypothesis, Selberg's paper stands as a template for success. For this reason, it has become one of the most studied papers in recent mathematical history. If only this could be emulated for the Riemann zeta function!

A near hyperbolic hustle

In 1977 it almost happened. In that year, H. Haas, a German master's student, took it upon himself to compute eigenvalues for some of the folded-over hyperbolic domains that had so intrigued Selberg. A list of numbers was generated, and it found its way to the desk of a number theorist at MIT, Harold Stark. As Stark browsed the masters thesis, which included a list of the eigenvalues, he did a double take, for within the list he saw numbers which many a mathematician has committed to memory: the first several zeros of the zeta function. Could it be that the zeta zeros were indeed to be discovered in hyperbolic space?

Stark quickly announced the discovery to the mathematical community, and soon a hunt was under way to understand the source of Haas's numbers. At the head of the investigation was Dennis Hejhal, then an assistant professor at Harvard University. After several months of work Hejhal determined that in fact the coincidence of zeros and eigenvalues had a rather mundane explanation. His detailed analysis of Haas's computational procedure revealed a subtle mistake in the eigenvalue calculation, one which produced zeta zeros as a computational artifact.

Back to the game room

Physicists continued to investigate these hyperbolic triangles closely. In particular, in the 1980s Bohigas and his colleagues found much evidence in support of the basic quantum chaos conjecture. But this work also showed that life in the Poincaré disk was not all chaos. It seemed possible to make hyperbolic tables that produced neither the randomness of Poisson nor the structure of GUE. The closer they looked, the curiouser and curiouser things appeared.

The surprising numerical experiments of Bohigas and miscalculations of Haas served as modern warnings against overreliance on calculation. Where are the clean, clear, beautiful theorems? If the road to the Riemann hypothesis were to go through hyperbolic space, it was clearly time to get the mathematicians more involved.

13 *Making Order Out of (Quantum) Chaos*

DESPITE the mounting numerical and graphical evidence for a connection between quantum chaos and the Riemann hypothesis, and despite the analogies to be found between prime geodesics and prime numbers, there was still no logically sound evidence that this road through randomness would lead to a Riemannian resolution. At best the experiments had opened a window on a new and exciting area at the border between mathematics and physics, which might shed new light on this old, difficult problem. At worst, a new sort of pictorial numerology had been created in which the spirit of Riemann shimmered in the energy levels of Hamiltonians.

Haas's false eigenvalues, like Skewe's number, provided the latest warning of the limits of computation, reminding us that numerical experimentation can go only so far in the search for absolute truth. Eventually we need to sit down to prove some theorems. Among the mathematicians leading this call to cogitation there is one who brings to bear a capacity for deep concentration that he developed as a child chess champion and another who works with the informed skepticism befitting a magician turned mathematician. Other mathematicians in the hunt are going so far as to create entire mathematical worlds outside the realm of physical law, where energy and prime number become one, in the hope that in doing so they might be building a final resting place for the Riemann hypothesis. These are some of the big players now in the game, and we'll meet them here.

A GRAND MASTER OF THE RIEMANN HYPOTHESIS

When the quantum chaos contingent realized that they would need to better understand the wave functions that they had set resonating on the hyperbolic drums, they turned to an expert, Peter Sarnak, professor of mathematics at Princeton University. Sarnak is by all accounts the unofficial world leader in research related to the Riemann hypothesis, but when the physicists approached him, it was primarily for his deep knowledge of ***spectral theory.*** This is the subject into which the investigations of matrices and eigenvalues have developed. Of particular relevance was Sarnak's expertise in dealing with matrices that encode the transformations of the continually negatively curved hills of hyperbolic spaces.

Sarnak brings a ferociously competitive nature to his pursuit of mathematical truth, an outlook derived from a childhood in Johannesburg, South Africa, mainly spent at the chessboard. From the ages of eight to fifteen, he played chess six or seven hours a day; ultimately, he became one of the best players in the country. He thrived on the gamesmanship but eventually decided to trade in the chessboard for the blackboard, having come to see mathematics as a source of far deeper and more interesting intellectual challenges.

These days Sarnak is a veritable academic quantum mechanical system, with a wave function that is spread around the leading mathematical institutions of the world. He has supervised many Ph.D.s, and as one student after another graduates he once again acts as the masterful chess player, looking ahead many moves at a time, positioning the pieces—students, collaborators, and ideas—in a well-orchestrated multipronged attack on the Riemann hypothesis.

An (Il-)Logical Road to Riemann

Sarnak's love of chess and the infinite complexity embedded in its relatively simple set of rules led to the study of mathematical logic, a subject in which surprises and beauty emerge from the manipulation of its own few symbolic pieces according to a spare Aristotlean rule book. So when it came time for Sarnak to go to graduate school, he chose Stanford University, in order to work with the eminent logician Paul J. Cohen (b. 1934).

Cohen received a Fields Medal in 1966 for his work in mathematical logic related to the first of Hilbert's twenty-three problems: the resolution of Cantor's continuum hypothesis, first stated in 1878. This is a problem that cuts to the heart of the mystery of the transfinite numbers—asking what number, if any, comes in between that number which quantifies the countability of the natural numbers, and that which stands in for the uncountable number of the real numbers. Cantor had asserted, but never proved, that there was no number between them, just as there is no natural number between two and three, i.e., that the number of the *continuum* immediately follows the number of the natural numbers.

Legend has it that Cohen, trained not in logic but in *harmonic analysis* (the descendant of Fourier analysis), solved the almost century-old problem in just several months, almost on a dare, the result of a Cardano-like challenge in which he asked a logician colleague to stump him with the most difficult problem in mathematical logic. Whereas Cardano had invented the imaginary numbers in order to untangle the intricacies of solving polynomial equations, Cohen invented the important technique of *forcing* in order to attack the continuum hypothesis. Rather than demonstrate either the truth or the falsity of the continuum hypothesis, Cohen proved that it is *independent* of the usual axioms that undergird arithmetic. Just as non-Euclidean geometries could exist by virtue of the independence of Euclid's parallel postulate from his other four axioms of geometry, so too could there be a variety of systems of arithmetic, which are the same when restricted to considerations of our familiar finite realm (i.e., the world of addition, multiplication, and division of integers), but different when pushed to the transfinite realm. Some had transfinite numbers between the countable and uncountable; others didn't. This result remains one of the important achievements of modern mathematics, and Sarnak had come to Stanford to study logic at the feet of a master.

The transition from undergraduate to graduate studies can be tough. Students must make the difficult leap from doing homework to conducting research, from solving problems that have known solutions to trying to solve problems that may have no solutions. Not one to suffer fools gladly, if at all, Cohen could be a terror to these aspiring yet often unsure mathematicians. One famous story has Cohen grumbling, "I thought we agreed you understood that," when a graduate student posed a simple question after several minutes of nodding anxiously during one of Cohen's complicated explanations. Nevertheless, the confident Sarnak was not easily cowed. Cohen took him on as a student, but while Sarnak had planned to follow in Cohen's footsteps as a logician, Cohen had other intentions. He had decided to switch gears and begin the pursuit of that "other" unsettled hypothesis in Hilbert's list, the Riemann hypothesis, a conjecture that seemed closer in spirit to his original interests in harmonic analysis. And he would take Sarnak with him.

Cohen had made an inspired choice of topic for this student, as Sarnak seems perfectly suited to work on a problem that has withstood the test of time. The hours spent puzzling over the chessboard had honed Sarnak's concentration and also taught him the importance of patience in approaching a difficult problem. Sarnak has remarked, "Research mathematics is frustrating. If it's not frustrating you're probably tackling problems that are too easy." In the physical sciences, you might be in the lab each day doing experiments and taking measurements. By contrast, "in mathematics your steady state is you are stuck. But, of course, you hope that you do solve problems on an occasional basis and you have a very big high for a short while. So if you're the kind of person who needs a daily high then it's probably not the right subject." To this day he admits that "the kind of problems that attract me . . . are problems that are crucial to understanding the foundational issues, the basic facts of either geometry or number. . . . Solving them or even making progress on them will leave your mark on the subject. . . . The problems are usually very hard and so one is usually stuck."

Cohen had Sarnak look closely at Selberg's hyperbolic techniques, and in particular wanted him to become expert in the properties of the standing waves that could vibrate in these hyperbolic spaces. To this day Sarnak keeps Selberg's collected works nearby, reading them over and over again, "as a religious man reads the Bible."

Sarnak's doctoral dissertation gave a careful analysis of the resonances and energies that are characteristic of wave functions in these closed hyperbolic regions, and thus marked his entry into the world of *spectral geometry,* a subject in which he is now one of the world's experts. So when the quantum chaologists needed to better understand the waves on the Poincaré disk, they turned to Sarnak.

Mathematician among the Physicists

As Sarnak recalls it, when he was approached by these scientists, his initial reaction was, "What are these guys talking about?" Indeed, for better or worse, physicists, who are held accountable to nature, as opposed to pure logic, are not as obsessed with proof as mathematicians are. Physicists are often happy to follow their instincts and build up a chain of conclusions that may or may not have a rock-solid logical foundation. This is the difference between developing a theory and proving a theorem. It is a process that can speed progress, allowing leaps of intuition that may anticipate centuries of rigorous justification, but can just as easily lead one far down a road that turns out to be an intellectual dead end.

Musing on the difference between physicists and mathematicians, Sarnak remarks, "In science you become convinced of the truth by some instances of it . . . basic arguments leading to the answer, but you don't give a formal rigorous proof. If this were physics, and you checked the first twenty [zeta zeros] you would say, oh, yes, of course it [the Riemann hypothesis] is true, so I mean at some level you could say this is true. In our world this is true. In mathematics, one is after not only the truth but a proof that it's true." Smiling, he says, "There are physicists who have written tens of papers with the words 'Riemann Hypothesis' in the title. I've written only one. Now if I could only write one more, but *the* one . . ."

The more Sarnak learned about quantum chaos, the more he became intrigued by what appeared to be a "tantalizing" connection with Riemann. Sarnak believes that "so often, people don't look outside their own narrow area." Having taken to heart Selberg's careful warning that "there have been very few attempts at proving the Riemann hypothesis, because, simply, no one has ever had any really good idea for how to go about it," Sarnak takes as a guiding principle "that you need to look outside your field for new ideas, for inspiration. That's what is good about the quantum chaos world. I was attracted to the new and fresh ideas."

Eventually he was led to the bible of random matrix theory, M. L. Mehta's *Random Matrices.* This thick book has proved to be a treasure trove for modern mathematicians, and as Sarnak worked through it, he slowly came to see the power of the theory, which led him to an almost religious belief in the link between random matrices and zeta zeros suggested by Odlyzko's calculations. Sarnak now expects, albeit "at a very speculative level," that there is a natural interpretation of the zeros of any member of the "zoo of zeta functions" as the eigenvalues of some infinite-dimensional matrix encoding the transformation of a Hilbert space, and that there is a single such Hilbert space that works for the whole lot of them. In short, he would say that the zeros are "spectral in nature," terminology that comes from the interpretation of eigenvalues as frequencies of light. Spectral indeed, as they would be the ghostly emanations of the primal material of arithmetic, first conjectured by that dead king of mathematics, Hilbert.

The pursuit of this rigorous connection is not a research adventure that Sarnak has undertaken by himself; his ability to collaborate is one of his great strengths. Some fifty years ago, most mathematics papers had a single author, a fact consistent with the stereotype of the lonely scholar locked away in his or her garret with only pencil, paper, book, and lamp. Today, multiple authorship is more the rule than the exception. Nevertheless, collaboration in a subject like mathematics, which seems to take place mainly in the head of the person doing it, is still a bit mysterious to many people. Sarnak points out that mathematics, like all sciences, has become increasingly specialized, so that, often, the solution to a problem may require "putting together different technical skills, and you might have expertise in one area and another person might have expertise in a different area and only this combination can get you further." It is a generous model of scientists working side by side as opposed to one stepping on top of the other, an image that takes into account the simple joy of working with another person: "It often makes the thing much more fun. . . . You sort of feed off each other. You throw out an idea and you start writing on the blackboard and if you get stuck you're stuck together. . . . At least there are two of us stuck."

With a junior faculty member at Princeton, Zeev Rudnick (now a professor at Tel Aviv University), Sarnak pushed Montgomery's work forward to show even stronger evidence for the connection between eigenvalues and zeta zeros. Montgomery's work implied that if the Riemann hypothesis were true, then the pair correlation of the zeros and that of the eigenvalues of a random Hermitian matrix would be the same. Sarnak and Rudnick have shown that even more detailed measures of correlation would also be true. This sort of reasoning extends to the even more general *extended Riemann hypothesis* that is conjectured to rule over the class of zeta functions making up the Dirichlet L-functions (the generalizations of Riemann's zeta function that arise in the study of primes in arithmetic progressions). Following along a line of reasoning much like Montgomery's, Sarnak and his former student Michael Rubinstein obtained computational evidence that if all the nontrivial zeros of any Dirichlet L-series lay on the critical line, then there should be an agreement between the fine-scale correlation of the L-series zeros and the eigenvalues of some other class of random matrices.

In each of these situations Sarnak has identified an implicit symmetry, a certain class of transformations that leave unchanged the underlying geometry, like the rotation of a sphere that leaves in place the actual space occupied by the sphere (in other words it has the effect of moving the points within the sphere among themselves). Matrices that embody rotational symmetries are said to be *orthogonal,* meaning that their corresponding spatial transformations, while moving about points in space, at least leave unchanged the angles between the lines that they define. A next level of generalization is given by the adaptation of orthogonal matrices necessary to account for a space that admits the notion of complex numbers. These are *unitary* matrices.

These were already mentioned as the matrices giving rise to the Grand Unitary Ensemble (GUE). Sarnak conjectures that the type of symmetry at work in the case of the matrix which would settle the Riemann hypothesis is *symplectic,* an even more elaborate generalization of the notions "orthogonal" and "unitary," preserving a more complicated notion of distance characteristic of certain more exotic Riemannian spaces.

The Whole Ball of Wax

To date, Sarnak's most important collaboration related to the Riemann hypothesis is his work with Princeton colleague Nick Katz.

They are a well-matched pair, two strong intellects with similarly strong personalities, capable of incorporating each other's ideas quickly, and equally capable of withstanding their own sometimes sharp, but always honest, criticism of each other. Katz is known for his expertise in the detailed and complicated calculations associated with work on the Riemann hypothesis. His love of detail and his inclination toward careful attention are seemingly at odds with his penchant for relaxing by riding a motorcycle in the countryside near Princeton. Seeking to connect the world of random matrices to proven mathematical truths, Sarnak and Katz took on a reinvestigation of the proof of the Riemann hypothesis over function fields, the landmark work of André Weil and its later generalization by Weil's colleague at the Institute, the Fields medalist Pierre Deligne.

Weil's proof, as well as Deligne's further generalization, produces a direct interpretation of the accompanying zeta zeros as eigenvalues of a matrixlike transformation called Frobenius after its creator, Georg Frobenius (1849–1917).

By most accounts Frobenius was a man of acerbic personality and thus an apt successor to the similarly cranky Kronecker in Berlin. Frobenius is one of the founders of the analytic aspect of symmetry called *group representation theory.* This is the investigation of the systematic assignment of matrices to symmetry operations; it realizes the symmetries as certain spatial transformations. The particular symmetry that Weil exploited permits a mixing up of the points on a curve—like the "symmetry" of a beaded necklace that rearranges the beads but leaves them all on the original strand.

By considering curves of increasing complexity Sarnak and Katz were able to discover that both the zeta zeros and the eigenvalues of the Frobenius transformation satisfy GUE spacing laws. So in this case we have all the ingredients tied together in one nice package: a set of zeta zeros that have an interpretation as eigenvalues and also satisfy a local spacing distribution that comes from a class of random matrices, all of which underlies a Riemann hypothesis that is true. It is another piece of evidence, and possibly the most important one, for a connection between eigenvalues and zeta zeros, and hence between operators and the Riemann hypothesis. Is this the light at the end of Dyson, Montgomery, and Odlyzko's tunnel? Or is that just an oncoming train?

Is it all a coincidence?

Sarnak and Katz's research gives inviolable mathematical proof of a deep connection between random matrix theory and the Riemann hypothesis for function fields. Nevertheless, despite their work, as well as an ever-growing body of numerical evidence (see Figure 44), the skeptic might still ask: Does all this necessarily foretell an analogous relationship for the original Riemann hypothesis? If science could have but one slogan, it might very well be Occam's razor, the most famous epigram of William of Occam (or Ockham), recommending that "given two explanations of an event, choose the simpler." Really, how likely does it seem that we might connect primes to quantum phenomena or see the rhythm of number in the caroms of a non-Euclidean billiard ball? Might it not all just be a coincidence?

If there is any mathematician alive who could hope to answer that question, it would be Stanford University's Persi Diaconis (b. 1952). Diaconis is one of the most renowned probabilists and statisticians of modern times, but even more, he has combined formidable technical skills with a unique outlook and astounding intellectual curiosity to become one of the widest-ranging mathematicians of our day. His work spans the continuous and the discrete, the applied and the pure.

Diaconis's mathematical origins are almost legendary in the scientific community. Although he and his sister were enrolled by their parents in the Juilliard School, and thus set on a road to a musical career, Diaconis had another sort of performing in mind for himself: he had his heart set on becoming a magician.

Diaconis was a youthful fixture in a twilight vaudevillian world of magic that once had its center in midtown Manhattan. At a "Round Table" of magicians that used to meet in the Wurlitzer building on Forty-second Street, Diaconis received his first exposure to the mathematics embedded in magic. Among the regulars was Martin Gardner, famous for his popular mathematical puzzlers in *Scientific American.* Gardner took Diaconis under his wing and began to show him the connections between mathematics and magic. Gardner introduced him to the zeros and ones of binary numbers, and Diaconis quickly applied this new knowledge in his first trick, which Gardner then included in a column in *Scientific American.* A mathemagician was born.

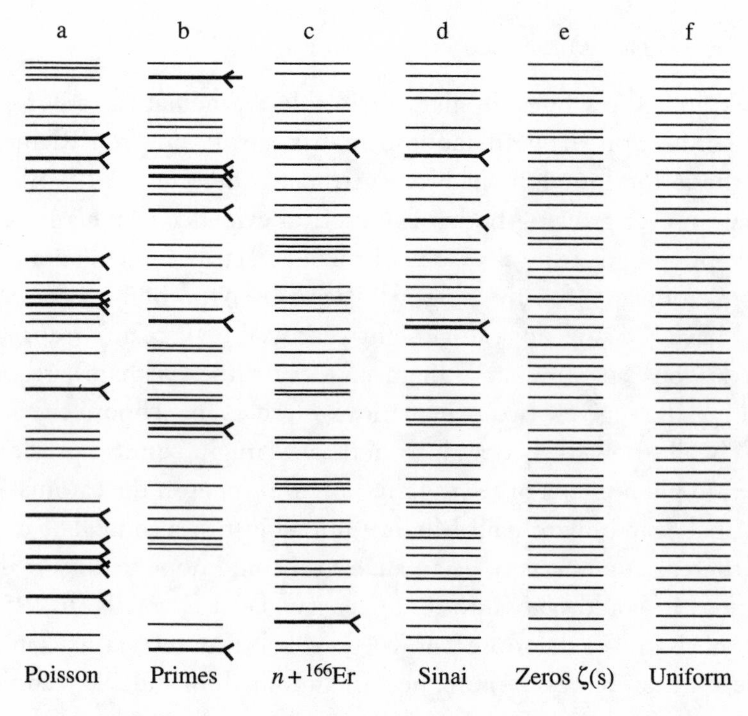

Figure 45. This figure, originally created by Bohigas and Giannoni, is among the most famous in the world of quantum-chaos-inspired research on the Riemann hypothesis. In column (a) are a set of lines thrown down at random positions along a line, dropped according to a Poisson process. Column (b) draws lines at the primes. Column (c) indicates a sequence of the energy levels high up in the spectrum of a particular isotope of the heavy element Erbium. Column (d) indicates the eigenvalues for the Hamiltonian which encodes the chaotic wanderings of a billiard ball constrained to the Sinai bumper pool table, a felt-filled rectangle with a circular central bumper. Column (e) marks off the positions of the first several zeta zeros, as they would appear on the critical line. Column (f) is a simple ladder of regularly or "uniformly" spaced segments. Notice the difference between the random lines in (a) and each of the other columns. The Poisson ladder has many more largish gaps and clumps. The zeros and Sinai billiards have a much greater tendency toward even distribution, a property often referred to as repulsion, for they seem to behave as though they were like-charged particles tossed along the line, repelling one another in such a way that the net effect is a nearly even distribution.

The mathematics that comes into play in magic is mainly ***discrete mathematics.*** Generally speaking this is the mathematics that concerns itself with the possibilities of the manipulations of finite collections of objects: for example, finding formulas to count the number of possible poker hands with two pairs, or the possible destinations of the top card in a deck after a sequence of shuffles or cuts of the cards. ***Combinatorics*** is the mathematics that derives wondrous formulas capable of enumerating the totalities of these and other sorts of complicated configurations. ***Group theory*** analyzes the possibilities of related manipulations. Gamblers and magicians who know these ideas can use the information to their advantage. This fascinated Diaconis, who recalls:

> *Mostly what I was doing in those days was hanging around with magicians trying to learn tricks. But there was this mathematical component of it that always fascinated me. It's combinatorics. There's been almost no magic trick I know of that uses calculus, for example. It's all manipulating finite things, shuffling cards or moving pennies around in a row or things like that, which is the type of mathematics I do now too.*

Having been bitten by the magic bug, Diaconis ran away from home to try to make his mark as a magician. He left at the age of fourteen to travel with the magician Dale Vernon. He spent the next ten adventure-filled years in picaresque wandering, performing in clubs and on cruise ships, becoming a master of "close-up magic," the sleight of hand that relies on little more than a deck of cards or a few coins, some fast talking, and even faster fingers. Magic is no different from any other art: the more you know, the more you want to know. Eventually Diaconis realized that in order to become an even better magician, he would need to learn more mathematics, and in particular *probability theory,* which is at the heart of many magic tricks. As Diaconis now tells it, this is what brought him back to academic mathematics:

> *I asked a friend what's the best book on probability that's an introductory book, and he told me William Feller's* An Introduction to

> Probability and Its Applications *and I bought a copy and I couldn't read it because I didn't know calculus. . . . I thought I was a smart kid; I thought I could do anything and well: I couldn't read that. So, I started college at twenty-four, to learn to read Feller's book.*

He finished City College in just two years, majoring in math. At the end of this time he decided that he wanted to continue his studies, and specifically that he wanted to go to Harvard University, less because of an intrinsic love of mathematics than simply to gain the respect of his family and peers. In spite of his rapid progress through the undergraduate curriculum, Diaconis was not a stellar student; his first three grades in calculus were two C's and one D. Harvard looked like a long shot. But mentor Martin Gardner came to the rescue. Although Diaconis's transcript and background might not impress the mathematics department at Harvard, his skill as a magician might be of interest to the department of statistics. So Gardner wrote a letter of recommendation saying, "Look, I don't know anything about math, but of the ten best card tricks invented in the last ten years this kid invented two of them and maybe you should let him in." One member of the graduate admissions committee in statistics was a friend of Gardner's, the mathematician and statistician Fred Mosteller. For Mosteller, Gardner's letter was reason enough to take a chance on Diaconis.

History shows that Mosteller made a good bet. Years later, Diaconis would return to Harvard as a tenured professor in the department of mathematics. He has been a recipient of many honors and awards, including a MacArthur "Genius" Fellowship (which he received in the second year of the award's existence) and is now a professor holding an endowed chair in both statistics and mathematics at Stanford University.

As Diaconis pursued his doctorate, his work on his dissertation began to touch upon some very deep analytic number theory. This forced him to find a second adviser in the department of mathematics. He managed to attract the attention of an up-and-coming junior faculty member, Dennis Hejhal, who later became famous for his trace formula and his calculations regarding the Riemann hypothesis.

It Takes a Trickster to Know a Trickster

Diaconis's background in magic has served him well scientifically, giving him a venue where he can meld skills of sleight of hand and sleight of mind. Many a magic trick relies on the seemingly serendipitous coincidence of two events: I think of a card, and you produce the same card. I think of a number, and you produce that number. The outcome is a surprise to the naive volunteer, but not to one well-versed in trickery and illusion. In one of Diaconis's most famous or (depending on what side you take in the controversy over parapsychology) infamous scientific feats he was called upon to observe a purported demonstration of extrasensory perception. His close observation of the demonstration by the presumed mentalist revealed that the results did not stand up to careful statistical analysis: they were no better than what would be expected by chance, after all the variability (or lack thereof) had been taken into account.

This sort of careful, skeptical thinking led Diaconis and Mosteller to examine the phenomenon of coincidence, which they defined as "a surprising occurrence of events, perceived as meaningfully related, with no apparent causal connection." The father of your new office mate shares your birthday; the license plate on the car ahead contains the first four digits of your phone number; a salesperson shares your last name—these sorts of events often give one pause. The psychiatrist Carl Jung attributed such events to a semimystical property of existence called synchronicity. Mosteller and Diaconis instead use statistical and probabilistic analyses to show that these events are not as unlikely as we might first imagine.

What about the Riemann hypothesis? How convincing is the evidence that the zeta zeros are like eigenvalues of a random matrix? Looks can be deceiving. Perhaps it is an instance of an effect which Diaconis and Mosteller called "the cost of close," a phenomenon typified by a feeling of coincidence that is just as likely to accompany an exact match of birthdays with a new acquaintance as a match which is simply within a few days or even within the same month—and for that matter within a few days of another acquaintance. We like coincidences, so we look for them and encourage them. The distributions for eigenvalues and their spacings are well understood and also well publicized (at least in the world of mathematics), so perhaps we're seeing a connection simply because we want to see one.

Diaconis had already published several important papers on aspects of random matrix theory when he decided to train his skeptic's eye on the frequently claimed connection between random matrices and the Riemann hypothesis. Motivated by the work of Jonathan Keating and Nina Snaith (one of Keating's students), Diaconis and a graduate student, Marc Coram, conducted a statistical "test of hypothesis" in order to quantify the degree to which the zeta zeros behave like the eigenvalues of a random matrix.

This sort of analysis is part of the world of *hypothesis testing* and *tests of significance,* techniques for quantifying the degree of surprise that should accompany an outcome in the face of a working assumption or hypothesis. It is a subject of humble beginnings, developed by the father of modern statistics, Sir R. A. Fisher (1890–1962), in order to judge if one of his colleagues could, as she had claimed, truly distinguish between tea into which milk had been poured, and milk into which tea had been poured. Fisher designed an experiment and analytic methodology which ultimately confirmed (beyond the shadow of statistical doubt) that this was the case. From these tea-tasting beginnings has grown a subject that now underlies all forms of statistical prediction, including the design and analysis of political polls as well as the investment decisions of hedge fund managers.

Diaconis and Coram used Keating and Snaith's correspondence between zeta zeros and eigenvalues and considered various test statistics computed from the zeta zeros. In particular they compared totals obtained by summing various collections of zeta zeros with those obtained by taking analogous sums of the eigenvalues of a random matrix. They found that the zeta zero sums are consistent with the hypothesis that the zeta zeros behave like eigenvalues, using the same sort of analysis as a pharmaceutical company that runs a trial to determine if a new drug is better than a placebo. This and other, finer-scale tests allowed them to conclude that overall the evidence shows "satisfying match-ups between the two domains."

Diaconis and Coram ended their analysis with a few more numerical tests, obtaining results that they describe as "tantalizing" but whose interpretation they "leave to the reader." Their Riemannian magic eight ball spins round and round and stops to reveal that "Signs Point to Yes," though one last spin leaves us with the words "Ask Again Later."

In search of a Hamiltonian to settle the hypothesis

Sarnak is among the first to say that in spite of all the exciting and beautiful mathematics and physics coming out of recent investigations into quantum chaos, it is "naive" to think that this work will culminate in the discovery of a physical system whose energies produce the zeta zeros, and thus a proof of the Riemann hypothesis.

Despite such pronouncements, the pursuit of physical roads to Riemann continue. Like detectives who doggedly work to get a better and better description of a suspect, Berry and his collaborators have gradually fine-tuned their characterization of the elusive Hamiltonian that would settle the Riemann hypothesis. Not only do chaotic dynamics seem necessary for its classical analogue, but Berry and Keating have determined that it must also lack *time-reversal symmetry.* In the classical setting, such a system would have the property that a viewer could distinguish between the backward and forward running of a movie of its evolution. In the quantum version this symmetry would be reflected in an analogous manner within the evolution of the probabilities of the configurations. Deeper analysis has succeeded in deriving increasingly detailed descriptions of the types of trajectories that would be discovered, and has even begun to quantify the extent of the chaotic behavior displayed by a system conjectured to settle the Riemann hypothesis.

It is for these reasons that, in Berry's eyes, the discovery of the physical system at the end of the Riemann hypothesis is potentially of such great importance. For should the Riemann hypothesis be confirmed in such a way as to make prophets of Hilbert and Pólya, then according to the basic conjectures of quantum chaos, as well as Bohr's correspondence principle, physicists would have at their disposal a well-studied example that would begin to shed light on the mysteries of the quantum-classical divide. What sort of classical system would appear at the semiclassical limit of this Riemannian Hamiltonian?

What sort of environment of particles and forces would be the classical paradigm whose infinitesimal analogue would produce a quantum system whose energy levels line up with the zeta zeros? For an integrable (i.e., nonchaotic) semiclassical limit, physics has the well-studied *harmonic oscillator,* realized by the motion of a mass at the end of a perfect spring or the vibration of a perfect guitar string, either of which has a quantization that is equally well understood. At the transition from chaos to quantum there is to date no such model, but Berry believes that were the Riemann hypothesis to be confirmed and "when (if) the operator . . . is found, it will surely be simple, and will provide a paradigm for quantum chaology comparable with the harmonic oscillator for quantum nonchaology."

In fact, Berry and Keating have even made a conjecture as to the nature of the Riemannian Hamiltonian, but they are as yet at a loss to explain the physical situation (i.e., geometry) which it will transform. One possible candidate is again a game of billiards, but one that would take place on a very, very tiny table in our mythical pool hall, the Chaotic Cue. This game of "neutrino billiards" would be played with the bouncing about of a nearly massless neutrino, one of the fundamental particles of nature, in the presence of a *four scalar potential,* a special external environment akin to the electromagnetic field generated near a high-tension wire or in a lightning storm. In addition, the Russian physicists A. F. Volkov and V. V. Pavlovskii have discovered that other possible candidates might be found among the collection of matrices describing the energetic configurations of electrons in semiconductors, the materials of microprocessors and memory chips. Ultimately, the secret to the Riemann hypothesis may lie not in the computations of Odlyzko and others, but in the physics of their actual computers. The medium very well could be the message.

IF YOU BUILD IT, THEY WILL COME: THE VISION OF A CONNES MAN

If there is no in vivo solution to the Riemann hypothesis, what about an in vitro solution? Perhaps we could simply create a matrix whose eigenvalues are the desired zeta zeros, and through studying this matrix, settle the Riemann hypothesis.

This has been the approach championed by the French Fields medalist Alain Connes (b. 1947), who is now a professor at the Collège de France. Connes received his Fields Medal in 1983, the ICM of 1982 having been postponed owing to the declaration of martial law in the host city, Warsaw, Poland. Connes was recognized for his work in the theory of *factors,* in essence, a deep investigation into a very general class of matrices related to quantum mechanics, which were first discovered and studied by John von Neumann.

Connes has cooked up a matrix whose eigenvalues are precisely the zeta zeros that lie on the critical line. In his interpretation this matrix describes a transformation in a very strange geometric world. It is wholly a mathematical creation, which, from the starting material of the primes, arrives not at the natural numbers dotting the single dimension of the number line but instead at a world of infinite dimensions.

The ***dimension*** of a space refers to how many numbers are needed to describe locations within it. We are comfortable with the limited lists of real numbers like the latitude-longitude pairs (two dimensions) that describe positions on earth; but Connes is working in a world of infinite dimensions where one dimension is reserved for a familiar real number, and each of the other dimensions draws from the strange worlds of *p-adic numbers.*

P-adic numbers are defined in terms of prime numbers. That's what puts the "p" in p-adic. So, for each prime in Connes's construction there is a single independent dimension of descriptors. The prime two provides a line of *2-adic* numbers from which to choose; the prime three provides a separate set of *3-adic* numbers; and so on. Even the one familiar dimension from which we draw only real numbers goes instead by the name of the dimension corresponding to the *infinite prime.* Real numbers permit us to describe phenomena with as much resolution as possible (e.g., think of a circle of radius one foot. If it is a perfect circle, its circumference would be equal to π feet; but since π has a never-ending decimal expansion, a perfect measurement, one of infinite resolution, would require all the power of the real numbers). However, each p-adic dimension provides only descriptors in which the resolution improves by powers of a given prime, as though it were a microscope or telescope that provides only magnifications of fixed powers at a given prime. For the 2-adic numbers, resolution improves by powers of two; for the 3-adics, it improves by powers of three; and so on. As the level of magnification improves, we see a growing, but always finite, collection of numbers, which in the infinite unattainable limit collapses into a continuum, but one in which the topology and arithmetic retain the vestiges of its finite origins. In this way the p-adics are both finite and infinite, discrete yet continuous.

The amalgam of these p-adic numbers and the real numbers is the abstract setting of the *adeles,* a space whose Joycean name hints at a world that is at once a mystery to intuition but amenable to formal mathematical technique. It is a perfect example of what Connes has called the "thought tools" that mathematicians invent for "the purpose of studying mathematical reality." For Connes this is a particular sort of reality populated by objects that have "an existence every bit as solid as external reality" but a reality "that cannot be located in space or time." But their study "affords—when one is fortunate enough to uncover the minutest portion of it—a sensation of extraordinary pleasure through the feeling of timelessness it produces."

Despite this avowedly Platonic outlook, Connes's adelic investigations are all part of his trailblazing inquiries into the physical world, accomplished with the tools of *noncommutative geometry.* This subject reinvents the Cartesian coordinate approach to geometry and replaces the single numbers that indicate aspects like length, width, and height—real numbers for which multiplication is ***commutative*** (i.e., if multiplied together they give the same result independent of the order in which they are multiplied)—with those basic mathematical creatures whose multiplication is ***noncommutative:*** matrices. This is a complicated mathematical environment, but possibly one which gives a footing for various string-theoretic models of the universe.

In this setting, the Riemann hypothesis is translated into a conjecture about the structure of a new and possibly equally mysterious adelic matrix. In fact, this is an approach that goes back at least as far as Sarnak's adviser, Cohen. To date, it is unclear if this reinterpretation is of any help at all, as it may have the effect of simply translating a mysterious text from one arcane language into another. Nevertheless, whereas Cohen has always for the most part kept his adelic ideas to himself, Connes has, to his credit, gone out on an intellectual limb, offering this approach to the rest of the scientific community as food for thought (as well as evaluation).

In the introduction to one of his best-known papers outlining his approach to Riemann, Connes recalls that his first teacher, Gustave Choquet, once remarked, "One does, by openly facing a well-known unsolved problem, run the risk of being remembered more by one's failure than anything else." Nevertheless, Connes concludes, "After reaching a certain age, I realized that waiting 'safely' until one reaches the end point of one's life is an equally self-defeating alternative."

Perhaps Choquet had in mind Einstein, who in the last decades of his life devoted himself to an ultimately fruitless search for a Grand Unified Theory that would reveal all the fundamental forces as instances of a single magnificent force. With its current position at the nexus of physics and mathematics, the resolution of the Riemann hypothesis may indeed point to a mathematical Grand Unified Theory. Grand aspirations require a measure of hubris, and if pride goeth before a fall, it almost always also accompanies great progress.

An Endgame Strategy for the Riemann Hypothesis?

Perhaps quantum chaos really is the bridge to the "good ideas" that Selberg knows will be needed to settle the Riemann hypothesis. But perhaps not. Will Berry be proved a knight-errant? Are the vibrations found at the semiclassical limit muffled snatches of a song sung by the zeros of the zeta function, or are they nanoscale whispers of a quixotic windmill? Has Sarnak reached the endgame of this Platonic chess match? Perhaps, working together, mathematicians and physicists might find a strategy to checkmate the Riemann hypothesis.

14 *God May Not Play Dice, but What about Cards?*

STANDING ON the shoulders of Dyson, Montgomery, and Odlyzko, mathematicians and physicists like Berry and Sarnak concentrated on the coarsest properties of the eigenvalues. Their research begins to yield an understanding of the energies of randomly arrayed quantum mechanical systems, but it is an understanding of structure that appears only when these energies are jumbled into the buckets of their differences—the distribution of the spacings. For this reason, an individual spacing distribution is often referred to as the *bulk spectrum.* Its striking similarity (in the case of a random Hermitian matrix) to the distribution of differences of the zeta zeros is the source of the hope that Pólya and Hilbert might have been right, that for some matrix the individual eigenvalues would match up exactly with the zeta zeros.

So we now proceed from spacings to specifics and seek a deeper understanding of the individual eigenvalues. If the resonances resident within a randomly chosen matrix hold the key to the Riemann hypothesis, we need to know more about them, not only as they apply to this hypothesis but in and of themselves as well. Like astronomers whose coarse understanding of the structure of a randomly chosen slice of the universe is but a first step toward understanding the particulars of the night sky, so do we now roll up our sleeves to delve into the detail of the spectrum of a random matrix, hoping to uncover the secret of the zeta zeros, one of the basic mysteries of the mathematical universe.

Thus it is fitting that an important insight was achieved, in the

1990s, by two teams of mathematicians who were striving to understand fundamental secrets of our physical universe. Craig Tracy and Harold Widom in California, and Jinho Baik, Kurt Johansson, and Percy Deift in New York City, discovered that in the specifics of the spectrum of a random matrix there exists a beautiful, even universal structure. Its explicatory reach extends from the quantum mechanical world of random matrices to the statistics of solitaire, from the dynamics of moving flame fronts to the waiting time in a bank queue. While working to understand aspects of a Grand Unified Theory that would knit together all physical phenomena in a single mathematical law, they uncovered a mathematical law that may unify all of mathematics around the Riemann hypothesis. We now take some time to walk along this edge of research into the Riemann hypothesis and meet some of the people working there.

A California collaboration

Craig Tracy was born in England (in 1945) but was raised in Kansas City, Missouri. His path to mathematics started with a Ph.D. from one of the world's top physics departments, at the State University of New York at Stony Brook. Tracy followed up this strong start with several postdoctoral fellowships, but despite a growing publication list, in 1976 he found himself on the verge of joblessness, unable to obtain a position in physics for the coming year.

Tracy's training and expertise were in *statistical physics,* the area that gave rise to Dyson's original investigations of the random matrix. While Tracy was attending an annual conference on statistical physics held at Rutgers University in New Jersey, his luck took a turn for the better. At some point in the meeting, where hundreds of discussions were taking place concerning particle collisions, Tracy was able to benefit from a human collision. He ran into the mathematical physicist Arthur Jaffe of Harvard. Jaffe knew of Tracy's work and, after learning of Tracy's trouble in securing a position for the coming year, suggested that he look at positions in mathematics departments. As Tracy recalls, this was one of the most important moments of his life. He applied to several places, and with the help of one last letter of recommendation from Jaffe was able to persuade the expert probabilist

Laurie Snell to hire him into Dartmouth's department of mathematics—despite the fact that Tracy had never taught a math course. He has never looked back, and he is currently a professor of mathematics at the University of California at Davis.

For Harold Widom, on the other hand, there never seemed to be much doubt that his future lay in mathematics. Widom, who was born in Newark, New Jersey (in 1932), and raised in Brooklyn, was one of those legendary math whiz kids. He graduated from a famous science hothouse, Stuyvesant High School, where he captained the mathematics team. Then, like his fellow New Yorker Diaconis, he enrolled at nearby City College of New York. Whereas Diaconis had a late start in college studies, Widom entered at the age of seventeen and after two years felt that he knew enough mathematics to begin graduate school. By this time he had already achieved a measure of mathematical notice as one of the five highest scorers in the famous Putnam Exam, a national mathematics competition whose winners regularly go on to become top mathematicians. The University of Chicago accepted the application of this confident prodigy and Widom enrolled there, leaving City College without a degree.

By age twenty-three Widom had earned his doctorate under the supervision of the renowned algebraist Irving Kaplansky and was on his way to becoming an expert in *operator theory,* a broad field that encompasses the world of matrices and is of crucial importance to quantum mechanics. Widom is now past seventy and an emeritus faculty member at the University of California at Santa Cruz. He is as active as ever, an iconoclast whose research record demolishes the myth that mathematicians over thirty cease to be creative.

Tracy and Widom usually lived and worked just a few hundred miles from each other, but when they began to collaborate they were an ocean apart. They were brought together in 1991 through Estelle Basor, a former student of Widom's who was a professor of mathematics at California Polytechnic State University at San Luis Obispo. Basor and Tracy had started a project related to some of the early work of Dyson, Gaudin, and Mehta on random matrices. These three had determined the probability that a random matrix does not have eigenvalues in a particular range. Tracy and Basor began to investigate the probability of finding one, two, three, or any fixed finite number of eigenvalues in a given range.

At this time Tracy was in Japan on sabbatical, and as he and Basor began to communicate (mainly by fax) about the problem, they soon found themselves stuck on a very gnarly technical point. Basor sensed that Widom would know how to work through the difficulty. Although usually the academic parent—the thesis adviser—introduces his or her academic children around the mathematical community, in this case the tables were turned. Basor contacted Widom, hoping that he might be interested in the question she and Tracy had raised, and Widom was eager to join them. Working together, they solved the problem, deriving the probability that a given number of eigenvalues fell within a particular range.

With this work, Basor, Tracy, and Widom made their first mark on the burgeoning field of random matrices. Moreover, Tracy and Widom had become very good friends. Basor had now become interested in other things, but Tracy and Widom decided to continue to work together. The bulk spectrum had proved interesting, but Tracy had heard rumors that other interesting things were happening somewhere at the edge of the spectrum, at the largest eigenvalues, so off they went to investigate. What they found was a mathematical behavior so nearly ubiquitous that it was akin to finding another kind of bell curve.

From a Grand Unification to a Grand Universality

The paradigm of mathematical universality is the bell curve, that wonderfully symmetric shape to which the quantification of so many phenomena seems inexorably drawn. Ask a large number of people their heights, graph the number of people with each height, and a bell curve emerges centered around the group average. Now do it with weights—once again, you get the same basic shape. And if we retooled the classic model of randomness, by having each of those famous million monkeys exchange their typewriters for a fair coin and had them each record the number of heads obtained in tossing that coin one million times, then a graph reflecting the collective work of these simian assistants, counting the number of monkeys whose experiment garnered a given number of heads, would also almost surely produce another bell-shaped curve.

The foregone nature of the bell curve is its defining property. It is the final resting place of the distribution of outcomes of any accumulation of repeated independent events, of which the successive tosses of a coin or rolls of dice are the classic examples. This is the guts of the classical ***central limit theorem,*** a more informative and textured version of the law of averages, which we have already met in the discussion of Stieltjes's random-walk-inspired assault on the Riemann hypothesis. The central limit theorem first appeared in raw form in 1718 in *The Doctrine of Chances,* a pamphlet written by the French mathematician Abraham de Moivre (1667–1754). Forced to emigrate during the expulsion of the Huguenots, de Moivre took up residence in England. He was a great friend of Newton's and was widely respected as an intellectual, but as a foreigner, he was never able to secure a teaching position. Legend has it that the recorded cause of de Moivre's death was "somnolence" and that he had predicted the date of his passing upon realizing that each day he was sleeping fifteen more minutes and therefore concluding that his life would end on the day when he slept for twenty-four hours. He saw his end as a steady accumulation of Edgar Allan Poe's tiny slices of death that are sleep, thereby proving a limit theorem for mortality.

The assumption of independence underlying de Moivre's discovery of the bell curve perhaps reveals his derivation as an artifact of the time, a simpler era when the world was less connected than it is today. What would be a bell curve for our own time? What would reflect a world characterized by six degrees of separation and 24/7 activity? Where is the universality in dependent events? Tracy had heard that it might be found in the recently discovered connections between random matrices and quantum gravity.

EIGENVALUES ON THE EDGE

The subject of quantum gravity is the attempt to find a mathematical formalism capable of unifying the seemingly disjointed physical phenomena of gravity and quantum mechanics. Newton invented the tools of calculus in order to investigate the consequences of his classical gravitational inverse square law, but today we realize that gravity is better understood in terms of Einstein's theory of general relativity.

Gravity is the last of the fundamental forces of nature to resist "unification," electricity, magnetism, and quantum mechanics having been knit together (with Dyson's help) into the Feynman-Schwinger-Tomonaga theory of quantum electrodynamics (QED), which was then incorporated into a theory of quantum chromodynamics (QCD) that could also account for the theories of the weak and strong nuclear forces. Gravity stands alone, an annoying but intriguing holdout to the creation of a Grand Unified Theory.

Through his study of quantum gravity Tracy had learned that the variation across the ensemble of either the highest energy or the second highest energy, or for that matter any fixed "highest" energy, gave evidence for a behavior that was as universal as the bell curve. This would be a universal behavior for dependent phenomena, as this largest energy is the summary of a complicated collection of interactions among the entries of a given matrix. These highest values at the top of the spectrum taken one at a time are the *edge* of the spectrum of a random matrix, and Tracy and Widom hoped to derive a formula explaining the new shape to which these distributions were attracted.

Tracy and Widom were aware of the work of a team of Japanese mathematicians based in Kyoto—M. Jimbo, T. Miwa, Y. Môri, and M. Sato—who had discovered a new path to Dyson and Wigner's findings on the bulk spectrum that measured the distribution of the eigenvalue differences. In a tail-biting tale of scientific research, the Kyoto school had derived its original inspiration from some of Tracy's earliest joint work in statistical physics. The most intriguing aspect of the Japanese mathematicians' results was a connection they made between the bulk spectrum and a certain class of partial differential equation solutions, the *Painlevé transcendants.*

The corresponding *Painlevé equations* are a family of differential equations, those terse indirect mathematical descriptions of phenomena in terms of their rates of change. The Painlevé equations are named for their discoverer, Paul Painlevé (1863–1933), a lifelong friend of Hadamard's, and they come in six flavors, unimaginatively called types I through VI.

Painlevé's geopolitical legacy is at least as noteworthy as his mathematical one. He entered French politics in 1906. During World War I he served as minister of war as well as premier, and in 1924 he was a presidential candidate. Painlevé's interest in and enthusiasm for aeronautics (he was Wilbur Wright's passenger in 1918) led to the creation of the French Air Ministry. He is famous for the foresight shown in his appointment of Pétain to the command of the French forces at Verdun, but perhaps infamous for his role as the moving force behind the construction of France's ill-conceived Maginot Line.

Fortunately for science, Painlevé's talents as a mathematician exceeded those as a military analyst, and the solutions to the Painlevé equations have recently resurfaced as a key concept in many areas of mathematics. These solutions take a place in the long line of important mathematical functions which are characterized as the solution to a certain differential equations. These include the well-known logarithm and exponential, as well as the familiar sine and cosine. Some of these solutions are called *special functions,* and they play an important role in describing many physical systems including weather, planetary motions, and medical imaging.

Painlevé was searching for a new way to generate special functions by using *nonlinear* differential equations, instead of the linear differential equations used by his predecessors. Nonlinear equations are notoriously hard to understand, and their unpredictability is the source of much of the difficulty in making accurate long-term weather forecasts. Painlevé studied only the simplest family of such equations: a collection of differential equations which, like the members of a biological family, share some basic form, but through the variation of a small number of parameters will yield different individuals. Eventually, driven mainly by aesthetic considerations, Painlevé and a colleague, B. Gambier, whittled down an original collection of fifty or so equations to a primordial set of six.

The Painlevé transcendants are the solutions to these equations, and they would begin to reveal their importance about seventy years later, in some of Tracy's joint work on the *Ising model.* Named for the physicist Ernst Ising (1900–1998), the Ising model is an infinite regular grid—in two dimensions looking like a city street map, and in three dimensions like a jungle gym—with the property that the intersections or junction points are inhabited by a single particle, which is assumed to exist in one of two *spin states,* often labeled as *up* and *down.*

Any particular *spin assignment* turns the lattice into a brindled map of regions of spins that are either all up and or all down. The typical questions asked of such a model are related to the regional patterns that can occur when a given initial spin assignment is allowed to evolve according to some *update law,* a specified rule for assigning a (possibly) new spin value to a particle, according to the current spin values of some subset of the other particles in the grid. As regards the Ising model, the update rule for a given particle usually depends only on the states of the neighboring lattice points. For example, a standard update rule is "majority rule," in which a particle is assigned the spin state currently occupied by the majority of its neighbors.

Over time, any initial configuration undergoes a process of transformation as the lattice of particles updates its spin assignment at each tick of the clock. Systems begin in a disordered arrangement, and we can ask questions about the long-term behavior of the system, especially as related to the degree of order or disorder that develops, at either long or short range. Versions of this dynamical system model the way magnetic or conductive properties of metals and other materials can change as the material is heated or cooled.

These models have implications not only for physical systems but for living and social systems as well. If instead, the particles represent tiny cardiac muscles, either contracted or not, the analysis lends insight into the synchronization necessary to keep our hearts pumping. If the particles stand in for people in a social network, then their spin states might reflect their political views, and the dynamics can reveal how a person's opinions evolve depending upon the opinions of others.

In the 1970s, Tracy had been among the mathematicians and physicists who made important progress in understanding how coordination and synchronization reveal themselves in the Ising model. This is the work that Jimbo's team had come across. They found further physical connections, this time to the fundamental particles of natures called *bosons,* subatomic particles akin to photons in that they are thought to transmit some of the fundamental forces of nature. The Painlevé transcendants used by the scientists in Kyoto would turn out to be intellectual bosons, capable of mediating mathematical thought across a range of disciplines. For upon receiving a fax from none other than Madan Lal Mehta (the same Mehta whose book *Random Matrices* had intrigued and inspired Sarnak) detailing a new solution to the problem that had inspired Jimbo and his group, Tracy and Widom were able to put all the pieces together and derive a formula for the distribution of the largest eigenvalue of a random matrix, the number representing the highest energy level of a random quantum system, in terms of Painlevé transcendant II. Using related techniques, they were then able to find the distribution for the second largest eigenvalue, and so on.

Over a series of papers, the first of which appeared in 1993, Tracy and Widom discovered that the distribution of the values of the largest eigenvalue gives rise to a shape that is a subtle, but still well-defined, variation of the perfectly symmetric bell curve. Moreover, this shape is but one of a whole family of nearly perfectly symmetric bumps, all of which together are now called *Tracy-Widom distributions.* Their derivation of the distribution in terms of the Painlevé transcendant opened wide the eyes of the scientific community, and scientists began to see Tracy-Widom distributions everywhere.

A NEW UNIVERSALITY IN TOWN: TRACY-WIDOM DISTRIBUTIONS

By producing a formula for the shape of this distribution, Tracy and Widom made it possible for a wide range of scientists to recognize similar or related phenomenology within their own work. Tracy-Widom distributions are now seen in everything from the distributions of service times in queues at the bank to various predicted fluctuations in semiconductor performance. They are being applied to finance and even computational biology. Although still relevant and useful in many working statistical models, the bell curve is a curve of past generations, who could still believe the myth of independent actors. The Tracy-Widom distributions are the shapes of our time, the shapes to which dependence ultimately must conform.

The extent to which so many different phenomena give rise to the Tracy-Widom distributions has remade the landscape of possibilities for universal behavior, leading one mathematician to remark:

> *We thought that the topography of the world of distributions was composed of gentle hills and valleys, maybe one valley constructed by the central limit theorem where all distributions that wandered*

> *nearby were sucked down toward the normal distribution. But now we see it as a much more diverse and varied place, with many new valleys in which we might forever disappear.*

These are the words of Percy Deift (b. 1945), a onetime chemical engineering student who is now a professor of mathematics at New York University. Deift would be largely responsible for showing the mathematical world that the dazzling discovery of Tracy and Widom was not a stray lump of intellectual gold, but instead the tip of a mathematical mother lode.

Education of a Mathematical Miner

Percy Deift had barely begun his undergraduate studies when he was bitten by the mathematics bug. The son of a shoe manufacturer, he was born into a nonacademic but intellectual household in Durban, South Africa, within an expatriate community of Jewish Latvian and Lithuanian refugees. Like many a scientifically talented child of immigrants, he had begun university studies intending to pursue a career in engineering. It was a sensible path, meant to provide the young man with a respectable profession, and one in which he would follow in the footsteps of his older brother. The skills he would acquire as an engineering student would be useful anywhere should there ever come another day when the family was forced to move.

Like other budding engineers at nearby Natal University, Deift enrolled in Math 1, a course required of every student whose studies might require some mathematics. Although oriented toward applications, Math 1 still provided its students with a strong theoretical background, going so far as to begin with an axiomatic construction of the real numbers à la Dedekind. This mode of abstract investigation was a revelation for Deift; it had (in his words) an "absolutely profound" effect on him. He now felt that he must study mathematics.

A changed man, Deift approached his father for permission to redirect his studies. In addition to Math 1, Deift had also taken Engineering 1 and done well, so his father asked him if it might not make sense to take one more year of engineering, just to make sure of his choice. Deift agreed and returned to continue with another year of engineering. Deift smiles as he recalls what happened at the end of that year. Still smitten by mathematics, he returned to his father and asked once more if he might change direction. But his father replied, "How can you change now, when you've gone this far?"

What next? Torn between filial and intellectual devotion, he found himself facing a fork in the road, and unknowingly following the advice of Yogi Berra, he took it. He finished his engineering degree but stayed an additional year in order to obtain a second undergraduate degree, in mathematics. He then decided to pursue a doctorate in engineering. On the surface this seemed like a reasonable compromise, allowing him to pursue mathematics, albeit in an engineering context.

Deift took up the study of mathematics related to a problem in mining, the chief industry of South Africa and the main focus of its chemical engineering business sector. In particular he studied the process of grinding, a technique used to extract copper from boulders containing ore. He soon discovered that engineers had little use for the theoretical abstractions into which he transformed their problems, and the mathematics that he was developing to analyze these problems was of little interest to mathematicians. Frustrated by this neither/nor intellectual state, Deift decided that he needed to return to the pure sciences and mathematics. Recognizing his unhappiness, his parents at last supported his decision.

Luck Equals Tenacity Plus Opportunity

Deift applied to graduate school at MIT and was rejected; he was told that he lacked sufficient training in mathematics and physics. In order to remedy this failing, he entered a masters program at a local institution, Rhodes University, and managed to arrange a research position with Professor Joseph Gledhill, famous for his derivations of models of the atmosphere of Jupiter. Over the year Deift learned enough mathematics and physics to be accepted into Princeton University's Program in Applied Mathematics.

At graduate school Deift was taken under the wing of the mathematical physicist Barry Simon, who introduced him to *scattering theory,* the mathematics used to infer an object's structure through analysis of the patterns created when it is bombarded with radiation. Scattering theory helps make visible the invisible on all scales, translating a reflected radio pulse into pictures of celestial objects, hospital X rays into road maps of the human skeleton, or a ping-ponging electron beam into a portrait of the DNA helix. It is more generally a part of the vast subject of *inverse problems,* whereby structure is inferred from indirect description, like a mathematical version of twenty questions.

From Princeton, Deift moved to the Courant Institute of Mathematical Sciences at New York University, where he was taken under the wing of Jürgen Moser (1928–1999). Moser was born in Königsberg (the town whose bridges led Euler to invent graph theory) and was for many years a professor of mathematics at Courant, even serving as its director for three years. Deift describes Moser as the sort of mathematician who would solve a problem so completely that when he was finished, it was "left for dead with all four feet in the air." This is a tenacity which he clearly passed on to his academic children.

Moser turned Deift on to the study of integrable systems—the relatively predictable dynamical systems that live across the way from chaos, typified by the motions of billiard balls on square and circular tables. In spite of their predictability, the detailed study of their long-term behavior as described by precise predictions of their trajectories remains an important and interesting field of study. Deift had found ways to bring his talents to bear on this field, utilizing an emerging description of integrable systems as instances of particular sorts of inverse problems called *Riemann-Hilbert problems,* a name which reflects their origins in yet another of Riemann's open problems from the study of complex numbers and geometry. Hilbert amended its statement slightly and then included it at number twenty-one among his "challenge" problems. When Deift decided to pursue the study of these problems with his old friend Xin Zhou (now at Duke University), it turned out to be blackjack. It would lead them to uncover the ubiquity of the Tracy-Widom distributions and thus take them to the edge of the Riemann hypothesis.

In looking back on his intellectual journey, Deift is amazed, seeing it as a serendipitous adventure in which "people just seemed to appear at crucial moments." Whereas Persi Diaconis might see this more as the "luck" that follows on the heels of hard work and an open mind, Percy Deift sees more of a larger design to events. Either way, some years later, a chance meeting would prove to be the one that made possible the last step in bridging the worlds of continuous and discrete mathematics in the hunt to settle the Riemann hypothesis.

A Kid in a Candy Store

Like Tracy, Deift also had gravitated toward the study of quantum gravity. Deift had his own tool kit, and he wanted to apply the Riemann-Hilbert techniques to some of the open problems in the area.

Deift and Zhou had pushed forward a 150-year-old technique, *steepest descent.* Its classical version provides a means of finding solutions to problems in a manner akin to the way water "solves the problem" of finding its way down a mountain by choosing the fall line. The classical version was useful for working with linear phenomena, but Deift and Zhou were interested in finding a version of steepest descent that would work with the unpredictable *nonlinear* phenomena like the butterfly wings of chaos. In these investigations Deift and Zhou were led to the development of a matrix version of steepest descent.

Working away at this problem and others, Deift began to see that among a great variety of ostensibly different problems, the same sort of behavior appeared when considered in the long run. Moreover, it was a behavior characterized in terms of random matrices.

This bell curve–like universality intrigued Deift, and he began to try to understand it. Like most mathematicians interested in random matrix theory, he turned to Mehta's *Random Matrices.* Echoing Sarnak, he reacted to it with a mixture of frustration and wonder, finding the experience of reading the book to be like "going to a candy store in the 1950s"—heaps of delicious-looking morsels, but no detailed labels explaining precisely what anything is.

But, like a persistent child hoping to cajole a candy out of the shop owner, Deift kept at it, and when in the summer of 1994 a conference focusing on random matrices was held at Mount Holyoke, he was in attendance. He and his family had decided to escape the heat of the city for upstate New York, and so Deift was commuting between their summer home and the college. When Friday rolled around, Deift began to make preparations to leave the meeting and drive back to spend the weekend with his family. As an observant Jew, he could not travel after sundown, and so he was worried that he might not have time to attend the last lecture of the day, given by Professor Kurt Johansson. Deift decided he'd go to the lecture and leave early if necessary. It was a decision that ultimately changed the course of his intellectual life.

Johansson had found a way to simultaneously relate the bell curve to random matrix theory as well as some of Deift's asymptotic results. Deift was riveted by Johansson's lecture. He remembers thinking, "Finally, someone is making mathematics out of Mehta's book!" Furthermore, it gave Deift hope that he might uncover a similar sort of behavior in his own research. Deift wanted Johansson to teach him everything he knew and so Deift invited Johansson to visit Courant in the coming academic year.

That fall Johansson came to Courant and gave a series of lectures. The information exchange was not a one-way street, for Johansson had also arrived with some questions which he was sure Deift would be able to help him answer. In particular, Johansson had in mind a problem related to *permutations,* the mathematician's abstraction of the card shuffle. As they settled into Deift's office to discuss these things, Deift thought to call in one his graduate students, Jinho Baik, to join the session. Baik had just cleared the last hurdle of graduate school before his dissertation, having recently finished his oral examinations. Deift could see how some of his own work would apply to Johansson's question and so thought that it might be just the thing for Baik to work on as a dissertation topic.

Over the course of the next year, the three worked away, connected by e-mail, until one day they arrived at the solution.

All along Deift had expected that they would solve the problem, but what he hadn't expected was the form of the answer. For at the end of their journey had appeared the Tracy-Widom eigenvalue distribution: the distribution that made sense of the largest eigenvalue of a random matrix (the highest energy level predicted by the dice throwing of quantum mechanics) also predicted the distribution of outcomes of a random game of solitaire. From shooting craps to dealing cards—no matter where in the heavenly casino we looked, we would find the Tracy-Widom distribution!

Deift soon telephoned Tracy, saying to him, "You'd better sit down, I'm going to tell you something really interesting." Although as a government official Painlevé had failed to design a barricade against the Germans, as a mathematician he had unwittingly built a bridge that would one day serve to connect the ordering of the eigenvalues to the disorder of a card shuffle, and Baik, Deift, and Johansson had discovered it. In doing so, they revealed to the entire mathematical world a previously hidden entrance to the Riemann hypothesis.

From Sorting to Shuffling

Many a mathematics student will first encounter permutations when investigating basic questions of *enumeration.* For example, how many ways can a deck of fifty-two cards be arranged? How many five-card hands are there in a deck of fifty-two cards? Answering these questions requires that we count or enumerate certain kinds of arrangements. The former is a question about permutations or orderings, the latter is a question about combinations. They both fall under the purview of *combinatorics.*

We've already met combinatorics, the subject that first intrigued the mathematically minded magician and skeptical inquirer Persi Diaconis. Combinatorics is to discrete mathematics as calculus is to the continuous realm of analysis. The goal of combinatorial analyses is to uncover clever formulas for counting complicated arrangements. Its origins can be traced back at least as far as 1800 B.C.E. in Egypt, where glimmers of the subject can be found in the historic Rhind Papyrus, a roll one foot wide by eighteen feet long bought in a resort town on the Nile in 1858 by a Scottish antiquarian, Henry Rhind.

The Rhind Papyrus is our chief source of information about the mathematical maturity of the ancient Egyptians. Among its many problems and solutions is one that echoes a classic children's puzzler involving a neighborhood where each of seven houses contains seven cats that each eat seven mice, each of which would have eaten seven ears of grain that would have produced seven measures of grain. The successive powers of seven that count the numbers of houses, cats, mice, and so on, can be deduced by basic techniques of enumeration, a fundamental tool of combinatorial analysis. Related techniques produce solutions to more advanced brain teasers such as, "How many people do you need at a party to ensure that at least three of the guests all know one another or at least three will not all know one another?" (six), or "How many people do you need in a room to ensure better than fifty-fifty odds that two share a birthday?" (at least twenty-three).

But for all the fun and games (which often provide an intellectual hook for budding mathematicians) the counting principles involved in answering such questions are also crucial to scientific considerations in a wide range of subjects.

For example, *combinatorial enumeration* helps the chemist navigate the vast landscapes of molecular possibilities used in drug design. Similar techniques have enabled the sequencing of the genome and are now playing a crucial role in gene identification. The scheduling problems that FedEx needs to solve in order to deliver packages to everyone everywhere overnight are dealt with by the techniques of *combinatorial optimization.* This application touches on the wide range of uses of combinatorics in computer science, which include analyzing the possible paths that our e-mail can take as it zips around the world and the clever data-compression schemes that enable the efficient storage of documents and images on your home computer.

In particular, the analysis and understanding of permutations appears to be central to many disciplines. Perhaps this is a reflection of the overarching goal of science as a pursuit of order in a disorderly world. Permutations arise in quantum mechanics as the symmetries of a physical system in which all the particles are considered interchangeable (i.e., in the search for those properties which are left unchanged after the system is "shuffled"). They arise in genetics and phylogeny in the study of the permutations the sequence of molecules making up a strand of DNA might experience during mutation over generations, or across a given generation. They are important in the design and formulation of statistical tests that require randomization. But it is safe to say that permutations are nowhere more important than in the subject of computer science.

One Person's Disorder Is Another Person's Order

Computers are mindless creations. They do exactly as they are told, forever executing automatically and without complaint one list of commands after another. A list of instructions, a recipe of logic designed to execute a simple mathematical or logical task, is an ***algorithm.***

The algorithm is the recipe, not the effect. A variety of algorithms having different advantages or disadvantages might accomplish the same task. There may be many ways to make a chocolate cake, and one procedure could be fastest, another tastiest, another cheapest. In the realm of computation, cake baking is not crucial, but *sorting* is tremendously important. Tasks such as ordering a list of numbers or alphabetizing a directory are frequently instrumental in the accomplishment of a much more complex goal. Data can be received in one order, but processing often requires that the data be rearranged into some new order. Demographic data may be entered alphabetically, but may then need to be ordered by income. Library lists may arrive alphabetized by title, but it might be better to order them according to author, and then within each author ordering according to date of publication. Sorting is a necessary step in preparing data for some larger computation. Such *preprocessing* is to a computing algorithm as is the prepping of ingredients to a cooking recipe.

Large computations can involve billions and even trillions of inputs. Tremendous efficiencies can be gained by a smart organization of the data flowing through the calculation. Such an organization or pattern of sorting ensures that any given number, physically existing as some pattern of magnetization on the hard drive, is located near the other data or microprocessors that it will need to encounter next along the computational path. For this reason, long hours can be spent outlining a route that will minimize the reorganization, or shuffling of the data, that is necessary.

This is much of the reason behind the long and symbiotic relationship between the mathematics of card shuffling and computer science. Jim Reeds, a computer scientist at AT&T, was responsible, in the 1960s, for the mathematical model of *riffle shuffling,* which attempts to mimic the way people really shuffle cards. Meanwhile,

Persi Diaconis's analysis of so-called *perfect shuffles* (an essential talent for any good cardsharp) had important implications for the design and understanding of networking schemes for parallel computer architecture. However, one of the first mathematicians to draw computational inspiration from a deck of cards was Stan Ulam, and it was in the answer to one simple question regarding an elementary card shuffle that a link between Painlevé and Riemann would appear.

A SUPERMathematiciAN

Stanislaw Ulam (1909–1984), a Polish-born mathematician, was part of the wave of eastern European émigré scientists who formed the core of the 1940s American nuclear weapons program. Although Edward Teller was indeed the driving force behind the development of the H-bomb, Ulam is surely its intellectual father, credited with the invention of the "staged process" (replacing an unworkable model of Teller's), which he and Teller eventually turned into the first working H-bomb design.

Ulam was originally brought to the United States in the 1930s by von Neumann, as a visitor to the Institute for Advanced Study, and chose to stay when World War II began. He became an American citizen in 1943. Soon thereafter, von Neumann drafted him to join the top-secret weapons lab at Los Alamos, as part of the small group already investigating the feasibility of a thermonuclear device (nicknamed "the Super") that would succeed the atomic bomb.

Ulam's primary responsibility was the mathematical modeling of various aspects of the bomb-making process that could not be investigated empirically for fear of a chance disastrous outcome. Ulam, along with Feynman, von Neumann, and Nicholas Metropolis, took on the challenge of programming the lumbering behemoths that were the first computers. They were looking for ingenious ways to organize the calculations that would be necessary to help simulate and predict the behavior of the atomic bomb.

Ulam was a man of great technical skills, but, unlike most mathematicians, he worked by talking rather than writing. Ulam was capable of carrying out extremely long, complicated computations in his head, a talent which some have attributed to a combination of terrible eyesight and a certain vanity that kept him from wearing glasses. Ulam's abilities derived from an amazing skill for finding great simplifications in complex calculations, which, when translated into directives to the nascent computers, would speed along the mechanical calculations.

The algorithms thus conceived focused on the evaluation of complicated mathematical formulas, as opposed to manipulations of lists of records or numbers. This defines them as falling within the purview of *numerical methods.* Here Ulam is most famous for his invention (conceived of jointly with von Neumann and Nicholas Metropolis) of the *Monte Carlo techniques,* which use randomness to speed up the knotty, time-intensive calculations necessary in weapons research, providing answers that, although not exact, are close enough to the precise answer to give a very good idea of the behavior under study.

According to Ulam, he arrived at the idea during the countless games of solitaire he played while recuperating from a nearly fatal bout of encephalitis. He began to realize that just from the layout of the first few cards, he could get a pretty good idea of his chances of winning. In the same way, he saw that the end result of many a calculation could be predicted by having some vague statistical understanding of the distribution of possible outcomes.

Today, any computational trick that uses randomness is called a Monte Carlo algorithm, and indeed, these are algorithms that in one way or another use an internal roulette wheel to generate the operational data for the computer during a calculation. Monte Carlo techniques are critical to the computer simulations that were necessary to test various versions of the atomic bomb, and even today are often used to replace testing by the actual detonation of a physical device.

Ulam brought randomness into calculation in the service of creating nuclear devices that might one day pitch the world into disarray, but it was in his study of sorting, the process of bringing *order* to calculation, that Ulam posed a problem whose solution would, in a sense, unify the mathematical world in the pursuit of the Riemann hypothesis.

Ulam's Problem

One familiar sorting algorithm formalizes the way many of us go about arranging a hand of cards. Sitting at the card table we receive our bridge hands. But of course we don't receive the cards in order. We are handed a mess, and then we try to make sense of it. Thirteen cards have been dealt to each of us, and we begin to arrange them. How to do this? We all have our own ideas, moving cards of the same suit close together, and then finally moving the cards about within the suits. Clubs over hearts, hearts over diamonds, and diamonds over spades. Within a suit cards are then ordered ace, king, and so on down to two. This is how we want to look at our hands, in order, and from this we design our bidding strategies and make our game plans.

Is there an optimal way of doing this?

It would be natural for Ulam, a computer scientist and expert bridge player, to think of a smart way to organize the cards in his hand. The basic ordering move of the cards is a *cycle.* Think of the cards as spread out in a line: a card is chosen, removed, and then reinserted somewhere else in the line, causing the intervening cards to be shifted either left or right by one move.

Any card player knows that any hand of cards can be ordered by performing some sequence of cycles, but how to do this optimally? Simply randomly cycling the cards around is no good, for we run the risk of going through every possible hand before coming to the ordered hand. How many possible ways of ordering are there? The first card could be any of the thirteen, the second card any of the remaining twelve, and so on. . . . All told this means that there are

$$13 \times 12 \times 11 \times 10 \times 9 \times 8 \times 7 \times 6 \times 5 \times 4 \times 3 \times 2 \times 1$$

possible orderings of these particular thirteen cards. This number, which is over 6 billion, is also called thirteen *factorial,* and is written, in a notation created by Euler, as 13! Factorials grow very quickly (perhaps that is the reason for using the exclamation mark). Thirteen factorial is already large enough so that every person in the world could hold any particular thirteen-card hand in a different order. Fifty-two factorial is so large that every person who ever lived could hold a deck of cards in a different order—and there would still be plenty of orders that were not chosen. (In fact, fifty-two factorial is even greater than the estimated number of protons in the universe.) Surely this is a number so large that we'd rather not just randomly cycle cards about, hoping for things to fall into place.

The minimum number of cycles needed to order the hand is called the *Ulam distance* between the given hand and the ordered hand. The less orderly the deck is when it arrives, the more time needed to rearrange it. One measure of this initial disorder is the length of the longest *rising sequence* in a permutation.

A rising sequence in an ordered list of cards is any (ordered) subset of the cards that are already in the correct relative order. For example, let's suppose that we receive a bridge hand in the following order (for simplification we'll assume that the cards are labeled one through thirteen):

2, 5, 1, 13, 7, 6, 11, 4, 10, 9, 3, 8, 12

There are many rising sequences within the sequence above. For example, we could start with 2, then go to 5, and then we need to ignore 1 (since it is less than 5) and then tack on 13. There are no numbers greater than 13 in the list above, so this sequence 2, 5, 13, cannot continue to increase. It is a rising sequence of length three. If we had chosen to ignore 13, we could have instead arrived at a rising sequence of length five: 2, 5, 6, 8, 12. Here, it is not too hard to check that there are no rising sequences of more than five numbers. So it is not only a rising sequence but is even a *longest* rising sequence.

The analysis of rising sequences is a good way of measuring disorder, for we can use the longest rising sequence as a heuristic for ordering the hand by cycles. Since the final ordered hand will maintain the relative order of these cards, we simply move the cards not included in this rising sequence, one at a time, from lowest to highest, to their appropriate positions. Figure 46 illustrates this idea by showing a sequence of cycles that sorts into increasing order the hand received as 2, 5, 1, 13, 7, 6, 11, 4, 10, 9, 3, 8, 12. The successive rows, from top to bottom, indicate the progression of the hand from disorder to order. At each step, the card just moved is indicated by boldface, with an arrow above it to show in which direction it was moved. For example, the boldfaced 1 leading the second row with an arrow above pointing left indicates that the 1 was moved from its position in the previous row to its current position by cycling it to the left.

2	5	1	13	7	6	11	4	10	9	3	8	12
←**1**	2	5	13	7	6	11	4	10	9	3	8	12
1	2	←**3**	5	13	7	6	11	4	10	9	8	12
1	2	3	←**4**	5	13	7	6	11	10	9	8	12
1	2	3	4	5	7	6	11	10	9	8	12	→**13**
1	2	3	4	5	6	→**7**	11	10	9	8	12	13
1	2	3	4	5	6	7	10	9	8	→**11**	12	13
1	2	3	4	5	6	7	9	8	→**10**	11	12	13
1	2	3	4	5	6	7	8	→**9**	10	11	12	13

Figure 46. A hand of cards (labeled 1 through 13) is put into increasing order using only cycles to rearrange the hand. At each step, the card just moved is indicated by boldface, with an arrow above it to show in which direction it was moved.

Ulam was able to show that the number of cycles needed to sort a bridge hand of any length is equal to the size of the hand less the length of the longest rising sequence. As we see in Figure 46, for this list of thirteen cards with a longest rising sequence of length five, eight cycles (thirteen minus five) were needed to put the hand in order.

Patience Sorting

It may have been solitaire that precipitated Ulam's dreams of Monte Carlo algorithms, but had he been playing the right kind of solitaire game he might have discovered the importance of the rising sequence even earlier. The length of the longest rising sequence also emerges as a measure of the outcome of a particular solitaire game, *Floyd's game* (named after another computer scientist, the Turing Award winner Robert Floyd). It is also called *patience sorting* because of its utility as a sorting algorithm, as well as its relation to a version of a game of patience (a Britishism for solitaire).

It is easiest to illustrate the game with a shuffled deck of cards, labeled with the numbers one through ten. The first card is turned over. This initiates pile one. For each succeeding card, if it is lower than any of the top cards (visible), put it on top of the leftmost one. If it is greater than all visible cards, it becomes the first card in a new pile.

For example, start with the disordered list 7, 2, 8, 1, 3, 4, 10, 6, 9, 5. Patience sorting creates piles in the following sequence shown in Figure 47 (the boldfaced numbers give each of the new insertions):

7

2
7

2
7 **8**

1
2
7 8

1
2 **3**
7 8

1
2 3
7 8 **4**

1
2 3
7 8 4 **10**

1
2 3 **6**
7 8 4 10

1
2 3 6
7 8 4 10 **9**

1 **5**
2 3 6
7 8 4 10 9

Figure 47. Evolution of piles created by patience sorting with initial input of cards labeled 7, 2, 8, 1, 3, 4, 10, 6, 9, 5.

Things now look even more disorganized, but then we perform the magic. We can remove the cards one by one, at each point removing the lowest top card, and, lo and behold, the cards are picked off in order. In a game of patience, the rule of placing the new card on the leftmost pile with a higher face card is relaxed to allow the player to place the new card on any pile showing a card of higher value.

According to the rules of patience the cards in any increasing subsequence in the original disorderly deck necessarily would have occupied locations in distinct piles (since they could never appear atop one another). Thus, the number of piles generated by patience sorting is indeed equal to the length of the longest rising sequence in the original.

In the Long Run

We now know for a particular disordered list, or hand of cards, how many moves we need to place the cards in order. But the cards or data really can come to us in any way, each new hand requiring a collection of moves that depends on the length of its longest rising sequence. These lists of data can be quite large: for example, think of the accounting information of McDonalds Corporation or the United States Census. So what any good computer scientist might want to know is how many moves, on average, we would expect to need to order the list. In other words, what are the asymptotics of the length of the longest rising sequence?

Starting in the 1960s, the analysis of algorithms became big business, for it was not until then that enough data were readily available to make this sort of analysis important. Also, in the 1960s analog-to-digital converters were invented. These machines took the continuous data of the world, often in the form of electrical signals, and sampled it, taking millions of "snapshots" of the data each second—just like a camera that very quickly takes a sequence of pictures of a person walking down the street. The sequence of snapshots is a digital representation of the continuous motion of the pedestrian. These converters thus had the ability to generate megabytes of data, an ability that previously had been very rare. This was a paradigm shift which marked a new standard size for a computational problem, and it had the concomitant effect of making necessary the analysis of the asymptotics of the algorithms. Through this sort of theoretical analysis it was possible to predict the expected performance of an algorithm when it was set to work on huge quantities of data. Today, asymptotic analysis remains an important part of computer science.

Recall that Ulam had helped invent Monte Carlo techniques in order to study the chaos of nuclear explosions. Now he used these same techniques to take on the challenge of bringing some order to the understanding of the average number of rising sequences in a permutation. He found that the ratio of the expected length of the longest rising sequence divided by the square root of the size of the hand approached a limiting value as the size of the hand increased. The exact calculation of this constant (determined independently by a two researchers at Bell Labs—B. Logan and L. Shepp—as well as by the Russian mathematicians S. Kerov and A. Vershik) came to be known as *Ulam's problem.* This result was of great interest to the computer science and combinatorics communities, and when some sixteen years later Deift, Baik, and Johansson discovered the full distribution of possibilities for the lengths of these rising sequences, revealed through the Painlevé transcendants, the entire mathematical world took notice.

A MYSTICAL CONNECTION

Card shuffling and solitaire are fun, but Baik, Deift, and Johansson had something else in mind when they decided to turn their attention to the problem of understanding the asymptotics of the longest rising sequence. In the 1960s, mathematicians had discovered an astounding connection between the mathematics of permutations and a mysterious game mixing combinatorics and geometry. In a process that almost seems pulled from the artist Sol Lewitt's notebooks, we start with a permutation and use this to form a pair of rectilinear arrangements of numerically labeled boxes according to a well-defined algorithm. These numerically infused shapes are called *Young tableaux,* in honor of the mathematician Alfred E. Young (1873–1940), who first used them to study permutations. For example, the permutation (of the ordered list of the first ten numbers) 7, 2, 8, 1, 3, 4, 10, 6, 9, 5 gives rise to the pair of Young tableaux in Figure 48.

1	3	6	7	9
2	5	8		
4	10			

1	3	4	5	9
2	6	10		
7	8			

Figure 48. A pair of standard Young tableaux, each of the same "shape" as dictated by the number of rows (three) and the length of each row (five, three, and two), as well as the fact that the numbers in their boxes appear in increasing order as you look across and also when you look down a given column. How these patterns are generated from a permutation is a bit involved; suffice it to say that the analysis of this process arises in almost every corner of mathematics, and its ubiquity remains a tantalizing and intriguing scientific puzzle.

The algorithm that performs the translation from permutation to sculpture-pair is the *Robinson-Schensted-Knuth* (RSK) construction. It is named for a mathematician at the University of Birmingham (England), Gilbert de Beauregard Robinson (who is an expert in the study of matrix-encoded symmetry that is group representation theory); the physicist Craige Schensted, formerly of the University of Michigan; and the computer scientist Donald E. Knuth of Stanford University.

Knuth is probably the most famous of the three. He is a recipient of almost every award that can be given to a computer scientist. In particular, he was honored with the Turing Award of 1974 for his contributions to the theory of algorithms and the Kyoto Prize of 1996 for advanced technology—the Japanese equivalent of the Nobel Prize.

Possibly, Knuth's most lasting contribution to mathematics and computer science is his creation and development of the word processing program and programming language *TeX,* built on his font design package METAFONT, which is now the de facto standard for papers written by mathematicians and computer scientists. The wild symbols and highly complicated notation and diagrams of mathematics can be a typesetting nightmare, and they are unmanageable for any standard word-processing package. Knuth decided to create *TeX* after becoming irritated with the results of the computer-aided typesetting put to use in producing the second volume of his three-volume masterpiece, *The Art of Computer Programming.*

For years mathematicians had found all sorts of strange and seemingly serendipitous connections between RSK and almost every field of mathematics. That this formal algorithm which attaches labeled shapes to permutations should affect so many different areas is astounding. It brings to mind Cayley's introduction of the formalism of matrices. Although matrices were introduced simply as an organizational tool to study systems of linear equations, the observation that a matrix could stand alone as a numberlike entity, with its own versions of multiplication and addition, remade mathematics on the spot. Unfortunately, though, RSK is so poorly understood that making any use of its connections has proved tremendously difficult. Finding a link to RSK was often an end point, albeit a familiar one, rather than any sort of beginning. RSK is so puzzling that mathematicians have barely been able to phrase a question that might help frame the study of its centrality, and this is perhaps one reason that Schensted has in recent years abandoned mathematics for a life of mystical inquiry.

It was already well known in the mathematical community that the length of the longest rising sequence in a permutation was equal to the length of the first row of its corresponding Young tableaux. Baik, Deift, and Johansson had a feeling that perhaps the long-term behavior of the other rows might be related to the other eigenvalues of a random matrix. Using their newly devised tools, they were able to produce a distribution that they believed would have to be the distribution of the length of the second row in the Young tableaux.

It is often easier to prove a theorem you already believe to be true, and so Baik, Deift, and Johansson sought some numerical evidence. They asked Odlyzko and another researcher at AT&T, Eric Rains (now a professor of mathematics alongside Tracy at UC Davis), to perform some calculations to see if the distribution they had derived agreed with experiment, and almost as if by magic, it did. Armed with this numerical confirmation, they were able to prove a theorem showing that the asymptotics of the distribution of the largest eigenvalue in a random matrix agreed with the asymptotics of the longest row in the RSK shape. They then conjectured a similar relation between the second row and second largest eigenvalue. In subsequent years this connection and further connections between the asymptotics of row lengths and eigenvalues have been proven.

These results gave indisputable evidence of a connection between random matrices and permutations, and thus between random matrices and the RSK algorithm. The RSK construction was developed by a mathematician, a physicist, and a computer scientist, and as befits its hybrid origins, it was revealed as a portal of discovery to what seems to be the entire mathematical world.

With these discoveries came an explosion of research activity. As Deift retells it, the announcement of the connection between RSK and random matrices was like an invitation shouted out to a collection of eagerly awaiting merrymakers: "Let the party begin!" And so they came, from far and wide: topologists and group theorists, combinatoricists and analysts, applied mathematicians and mathematical physicists. At the head of the pack were those who stalk the Riemann hypothesis.

Does the connection between the zeta zeros and random matrices then devolve to a link between the zeta zeros and number-filled boxes? Is there a link between the primes and the playfully dependent process that is RSK? Stieltjes believed he had found a proof of Riemann's hypothesis based on the accumulated ups and downs of the Möbius function, which von Sterneck thought to compare to the backs and forths of a drunkard careening along an endless street, or an endless game of coin tossing. We still don't know if Stieltjes had discovered a random walk to a proof, but now we can wonder if instead the key to the Riemann hypothesis lies in this new game of chance. Could the zeta zeros be related to a cosmic card game of solitaire, a curious queue, or an RSK-like sculpture of squares? Will we, like Stieltjes, find a Möbius capable of turning the zeta function inside out in a new and different fashion? Will our age have a Stieltjes whose own von Sterneck will compute this new game to the limits of the computational power of the day?

Coming full circle

The work of Tracy, Widom, Deift, their colleagues, and their collaborators leaves us with new questions, but nevertheless we've made progress—from an understanding of eigenvalues shuffled into their differences to the beginnings of an understanding of order.

In this we hear a faint echo of the earliest days of number, when number was still both order and aggregate, the palest of shadows of a time before the abstraction of the calculi. For the quantification that is number grew from the qualification that characterizes distinction. Before the iconic pebbles there were the particulars, in which identity might take the form of ordering. Before there was a generic "four," there were specific instances of four, perhaps identified according to order of acquisition, or ordered lightest to heaviest, or darkest to palest. Each ordering could be seen as distinct, like an infant who sees a new plaything in each rearrangement of the same set of blocks.

So in the work of Tracy and Widom and Deift stirs a reunification of number. Our travels to and through the Riemann hypothesis seem to be bringing us full circle, and that circle encloses all of mathematics.

Pólya had said that if you can't solve a problem, change it into one you can solve. The collective work of Tracy and Widom, and Deift, Baik, and Johansson showed that it might be possible to change the Riemann hypothesis into a problem in almost any area of mathematics: one about the geometric deformations that are the concerns of topologists, one about the waiting times at a bank that are the concerns of probabilists or optimization theorists, one about complex numbers for the analysts, one about integers for the number theorists, one about bouncing billiard balls for the physicists, one about a vibrating drumhead for the geometers, or even one about shuffling cards. Fittingly, for these connections Deift received the Pólya Prize in 1998 from the Society of Industrial and Applied Mathematics (sharing it with Sarnak and Zhou), and Tracy and Widom shared the prize in 2002.

In this web of connections we truly see the stature of the Riemann hypothesis. A great problem of mathematics becomes an intellectual nexus, providing a bridge across subjects and connecting seemingly disparate ideas, thereby allowing the tools and understanding of one concept to advance another. Furthermore, such a problem is not only a conduit but also a magical mirror, wherein any subject can be seen afresh, newly illuminated to reveal heretofore hidden beauty. In this way, the great problems are universal, reaching all of mathematics, so that progress toward their solution is really mathematical progress at large, and conversely, as the body of mathematical knowledge grows, so too does our appreciation and understanding of these landmark problems mature. Thus, we see the foresight in Hilbert's response when he was asked what was the most important problem in mathematics: "The problem of the zeros of the zeta function. Not only in mathematics. But, absolutely most important."*

And finally, with its relevance to almost all of mathematics laid bare, most every mathematician can have a chance to touch and perchance to dream of contributing to, and (dare we say!) even settling, this most important open problem in mathematics that is the Riemann hypothesis.

*From Constance Reid's *Hilbert.*

15 *The Millennium Meeting*

This is the journey of ideas and people that arrived in New York City in the summer of 2002—a journey that had begun with those first great mathematicians Euclid and Pythagoras and the discovery of the original mathematical material that is number, an intellectual primordial soup from which has evolved the rich diversity of ideas that is modern mathematics. Like the accumulation of primes that we hope to explain, our story has moved forward in a sequence of uneven fits and starts, although at a distance it may give the appearance of a great and unwavering curve of intellectual progress.

Just a few months before the conference, IBM researcher Sebastian Wedeniwski announced that using a cluster of more than 500 computers, he had confirmed the Riemann hypothesis for the first 50 billion zeta zeros. Now, as the many mathematicians, wearing mainly casual chinos, tennis shirts, and sneakers, mill about the ground floor of the Courant Institute of Mathematical Sciences (CIMS), the hallway is filled with animated and intense discussions of random matrices and zeta functions, billiard balls and permutations.

A glance at the schedule shows a full docket of lectures. The titles are certainly inscrutable to the uninitiated, and in some cases even to participants who have persevered through years of graduate and postgraduate mathematical education. Of all the scheduled lectures, none are more mysterious than the few titled "TBA," short for "to be announced." This built-in indeterminacy is somewhat reminiscent of a meeting that took place just a lucky seven years before in Cambridge, England, at which the mathematician Andrew Wiles of

Princeton University surprised the mathematical world with a slow intellectual striptease that ended with the revelation of an unannounced proof of the famous 350-year-old problem, Fermat's Last Theorem. Might the apparently unassigned spots support a similar bit of theatricality? Might they be placeholders for another neatly orchestrated intellectual dance of veils, intended to reveal a closely guarded solution to the long-standing and now riches-rewarding Riemann hypothesis?

The meeting is finally called to order by Professor Charlie Newman. Just the night before, Newman was appointed CIMS's interim director. He is a good-natured, good-humored man and an excellent mathematician. Like Craig Tracy, he is an expert in the study of interacting particle systems, good training for someone in charge of the academic tower of Babel that is CIMS.

A quick scan of the room reveals most of the Riemannian suspects. Wiles is in attendance, sitting quietly amid a swirl of rumors that he has now set his sights on the Riemann hypothesis. It is not lost on those in attendance that Wiles's proof of Fermat made use of a deft analysis of hyperbolic geometry, the same geometry that applies to the quantum chaologists' billiard tables. Deift has come down from his upstairs office and moves to a seat. Sarnak settles into his usual position at the front of the lecture hall, from where he will rattle off questions; his collaborator Nick Katz is in a seat nearby. Although Berry has not made the trip, some of his disciples are there, including Nina Snaith, whose recent work with Keating has resulted in some of the most rigorous connections between Riemann's zeta function and quantum chaos. Neither is Connes in attendance, but Belgian mathematician Christoph Denninger is present and is scheduled to present his own related geometric approach to the Riemann hypothesis. To some, Denninger's complicated ideas seem at present the best bet for a way to a proof of the Riemann hypothesis. Also missing is the irascible Louis de Branges de Bourcia of Purdue University. This mathematician sent a small tremor through the mathematical world by announcing a proof of the Riemann hypothesis, written in the form of an "apology"; it closed with his promise to use the Clay Prize to restore his ruined family château in France for use as a mathematics research institute. Although de Branges is famous for his solution of the *Bieber-*

bach conjecture in 1984, he has acquired a reputation for announcing proofs of famous problems but accompanying these claims with little or incorrect justification. His claims regarding the Riemann hypothesis are of this nature, and aspects of them have been refuted in various research papers by other mathematicians. In any case, however, he is not at CIMS to stake any sort of claim.

Neither is Dyson present, but Montgomery is there, as are Odlyzko and Hejhal. Selberg is also in attendance; the éminence grise of the Riemann hypothesis, he has been lured from his home in Princeton, still looking for the "new idea" that he believes necessary for proving the Riemann hypothesis.

Newman is scheduled to give a lecture at the conference, but that will come later. For the moment he wears his newly acquired administrator's hat and he welcomes the standing-room-only crowd with a little math joke. He finishes his opening remarks and quickly hands the reins over to Sarnak. In contrast to the casual dress of most of the audience, Sarnak's clothing suggests a real downtown New Yorker: a charcoal-gray jacket, black shirt, and black pants. Sarnak reminds the crowd of the progress that has been made but points out that there is still much to be done. As Sarnak ends his opening remarks and reviews the schedule, he jokes that he did not bring a proof of the Riemann hypothesis and asks if perhaps anyone else did.

It seems that for a moment, a sharp quiet engulfs the hall. Nervous, excited glances are exchanged, as the audience looks to see if there is any hint of an Arthur among the masses, ready to step forward and answer Sarnak's call to seize the sword from the stone. But the moment passes with just a chuckle or two—this will not be a Fermat-like occasion of a glorious mathematical surprise. Sarnak turns to his transparencies and begins his lecture. The audience members settle back into their seats, digging in for the week's worth of lectures on the Riemann hypothesis and random matrices.

Epilogue

Since that meeting in 2002, research on primes and the Riemann hypothesis has continued at a rapid pace, reinvigorated by the sudden appearance of newfound connections to physics and to every corner of mathematics. Meanwhile, the exponential growth of electronic commerce and communications has given real-world urgency to the study and understanding of the primes.

A few months after the NYU meeting, in the fall of 2002, a team of three young computer scientists, Manindra Agrawal, Neeraj Kayal, and Nitin Saxena, of the Indian Institute of Technology in Kanpur (the "MIT of India"), astounded the world of mathematics and computer science with the announcement of a new, blazingly fast, yet incredibly simple algorithm for checking to see if a number is prime (i.e., testing primality). The efficiency of an algorithm is measured in terms of the number of steps it requires, considered as a function of the "size" of the input. In particular, the size of the input to an algorithm for testing primality is roughly the number of digits in the candidate number. Until Agrawal, Kayal, and Saxena's discovery, all known algorithms for testing primality required a number of steps that grew exponentially with the size of the input.* They discovered a simple and elegant *polynomial*-time algorithm, that is, an algorithm requiring a number of steps approximately equal to the twelfth power of the size of the input. This is still a large number, but one that doesn't grow anything like an exponential—and the algorithm is a new landmark in our understanding of computational number theory. With the exponential barrier now broken, mathematicians have found ways to improve upon Agrawal, Kayal, and Saxena's result, and today the state of the art is an algorithm invented by Hendrik Lenstra of UC Berkeley and Carl Pomerance of Dartmouth College. Their algorithm requires a number of operations that is approximately the sixth power of the number of digits of the candidate number.

*More precisely, all known *deterministic* algorithms were exponential. A deterministic algorithm is such that there is no indeterminacy in its output—if it says a number is (or is not) prime, then that is true. On the other hand, there are known efficient *probabilistic* algorithms, whose conclusions have some very small chance of being wrong. These algorithms generally work in a randomized fashion: that is, they contain steps at which, effectively, the computer "flips a coin" in order to decide what to do next. Thus if these probabilistic algorithms are repeated several times and always give the same answer, either always returning the conclusion that the number is prime, or always returning the conclusion that the number is composite, then the likelihood of making a mistake in drawing such a conclusion is very, very, very small. So probabilistic algorithms are still quite useful.

Our understanding of primes in an arithmetic progression has also taken a leap forward. We have already seen how the work of Dirichlet, Hadamard, and de la Vallée-Poussin showed that, with a few obvious exceptions, the primes were evenly distributed among arithmetic progressions of a fixed step size. A related question is how many consecutive entries in an arithmetic progression are prime. For example, the arithmetic sequence of step size 210, beginning with 199, starts off with nine primes in a row: 199, 409, 619, 829, 1039, 1249, 1459, 1669, 1879, and 2089. It is natural to wonder—and the question can be traced back to the late 1700s—whether there exist arbitrarily long sequences of primes in an arithmetic progression.

In the spring of 2004, Ben Green, a postdoctoral fellow at the Pacific Institute of Mathematical Sciences, and Terrence Tao, a professor of mathematics at UCLA, startled the mathematics world by answering this question in the affirmative. Their proof is *nonconstructive,* which is to say that it proves the existence of such arbitrarily long prime arithmetic progressions without actually specifying how to make them. Thus we have yet another instance of one question answered only to give rise to many new, unanswered questions.

Soon after Green and Tao's paper was circulated, number theory was rocked by yet another grand announcement: R. F. Arenstorf of Vanderbilt University (now the American home of Alain Connes) submitted a paper to the electronic publication archive at Los Alamos National Laboratory claiming to have proved the ***twin prime conjecture***—that there are an infinite number of pairs of primes, like the pair 17 and 19, so-called twin primes that differ by only two. Upon investigation, though, an error was discovered, and the paper was quickly (and quietly) withdrawn.

In the summer of 2004, there was another explosion on the Riemann hypothesis front. Through a nonstandard venue—a press release by Purdue University—and with his eye on the million-dollar award that had been offered for the first person announcing a proof, de Branges claimed once again that he had found a proof of the Riemann hypothesis. He made public a link to a 124-page paper that could be retrieved from the World Wide Web. The announcement generated a flurry of Internet activity as well as a smattering of articles and interviews in the popular press, but to date there appears to be little professional interest in the paper. The manuscript claims to link quantum mechanics and the zeta zeros, so at least in spirit it is like the work of Berry, Sarnak, Connes, and others, but it seems to have very little, if anything, to do with quantum mechanics. Nevertheless, perhaps it does have the kernel of the "new idea" that seems necessary for solving Riemann's tantalizing puzzle. Only time will tell.

All the while, as a steady-state background hum to these sporadic explosions and implosions of research, more than 10,000 computers tied together on the Internet are cooperating to steadily produce zeta zeros. This distributed computing initiative, the ZetaGrid project, was the brainchild of Sebastian Wedeniwski, and grew out of his initial use of cluster computing for zeta zeros. Like its more famous counterpart, the SETI project, which harnesses the spare power of computers around the world in a *Contact*-like search for extraterrestrial life, ZetaGrid is an international collaboration comprising more than 150 teams of professional researchers and amateur zeta zero zealots who allow their computers to run software that computes zeta zeros when the computer is idle. Using the Internet as its communications backbone, members of ZetaGrid make use of one another's home calculations in order to compute zeta zeros at increasingly astronomical heights of the critical strip. Having begun operation in August 2002 as an Internet activity, ZetaGrid has, to date, produced some 830 billion zeta zeros—all on the critical line, and thus all confirming the Riemann hypothesis.

Indeed, after almost 150 years, the Riemann hypothesis remains alive and well. This is mathematical big game that remains at large, a number theoretic Loch Ness monster, a missing link that might one day tie together physics and numbers. So, on with the hunt for its proof.

GLOSSARY

Absolute error. Given the truth and an approximation to the truth, the absolute error in the approximation is the magnitude of the difference between the two values.

Algebraic geometry. Subject that developed from Descartes's introduction of coordinates into geometry. This made it possible to use equations to define geometric objects: curves, lines, surfaces, etc. Algebraic geometry is the study of the interplay between geometry and equations.

Algebraic number. Number having the property that some combination of multiples of its powers sums to zero. For example, a square root of two is an algebraic number because squaring it (taking its second power) and adding it to -2 (i.e., adding -2 times the zeroth power of two) gives zero.

Algorithm. Sequence of instructions (used, for instance, by a computer) to accomplish a specific task (e.g., sorting a list of numbers).

Amplitude. Term used to describe one characteristic of a regular sinusoidal wave. Amplitude is the maximum deviation of such a wave from its rest position. See ***sinusoid.***

Analysis. Modern term for the subject that grew out of the ***calculus.*** Generally speaking it is the mathematics that studies infinitesimal change.

Arithmetic progression. Sequence of ***integers*** in which the differences between successive integers is a fixed number. For example, the sequence of odd numbers, 1, 3, 5, . . . is an arithmetic progression in which the difference between successive entries is two.

Asymptotics. Study of the behavior of mathematical functions at very large values of the input. Thought of as studying a ***function*** as the size of the input "goes to infinity."

Base. Used in the context of ***exponentiation.*** If you raise a number to a power, the number being raised to the power is the base.

Basic conjecture of quantum chaos. This conjecture concerns the ***spacing distribution*** for the energy levels for a ***quantum mechanical Hamiltonian.*** It asserts that the spacing distribution of a ***quantized*** classical system depends only on whether the ***classical dynamics*** are ***chaotic*** or ***integrable.*** The conjecture is that in the former case the spacing distribution is ***asymptotically*** like that for the ***eigenvalues*** of a random ***Hermitian matrix,*** and in the latter it is like the spacing distribution for numbers generated according to a ***Poisson process.*** The first part of this conjecture is due to Bohigas, Giannoni, and Schmit; the other half is due to Berry and Tabor. See also ***quantum mechanics.***

Binary expansion. Representation of any ***natural number*** as a sum of powers of two, using a sequence of zeros and ones that indicate if a power is used or omitted. E.g., 1101 is the ***binary expansion*** of $13 = (1 \times 2^3) + (1 \times 2^2) + (0 \times 2^1) + (1 \times 2^0)$.

Binary numbers. Finite sequences of zeros and ones that are interpreted as ***binary expansions.***

Bit. Unit of information or computer memory that is either zero or one.

Calculus. Mathematics developed by Newton and Leibniz to quantify infinitesimal change.

Cardinal number. Natural number that represents a quantity.

Cardinality. Number of objects in a list (or set of things).

Cartesian plane. The familiar "*x,y* plane" of high school geometry, named for its creator, Descartes.

Central limit theorem. According to this theorem in its most basic form, the manner in which the average (normalized) value of a sequence of independent random outcomes distributed asymptotically approaches a bell curve. See ***asymptotics.***

Chaos. Characteristics exhibited by chaotic ***dynamical systems.***

Chaotic dynamics. Categorization of classical dynamical systems generally characterized by their ***sensitive dependence on initial conditions.***

Classical dynamics, classical mechanics. Description of the motion of objects according to Newton's laws of motion. This results in a model for motion using a system of ***differential equations*** according to which a solution to this system describes the motion. This works very well for macroscopic systems—e.g., the orbits of planets or the trajectory of a thrown baseball. This is implicitly a deterministic viewpoint, whereby if you know the initial conditions and the equations of motion for the system, then in principle everything is known about the future of the state of the system.

Combinatorics. Mathematics concerned with the derivation of formulas and techniques for enumerating complicated arrangements (e.g., the number of ways of seating a given number of couples around a table so that no partners are seated next to each other, or the number of matrices of a given size with integer entries such that the rows and columns each add up to some previously specified numbers).

Commutative. Term referring to a mathematical operation, meaning that the outcome is independent of the order in which it is performed. For example, since for any two numbers A and B, $A \times B = B \times A$, multiplication is commutative.

Complex analysis. Extension of the ***calculus*** to include ***complex numbers.***

Complex numbers. Numbers obtained by tossing into the ***real numbers*** a square root of minus one, usually denoted as ***i.*** Complex numbers are written as $A + iB$ (e.g., $-1/2 + 6i$) where A and B are both ***real numbers,*** called the ***real part*** and ***imaginary part,*** respectively.

Complex plane. Geometric representation of the ***complex numbers*** using the ***real parts*** as the *x*-coordinates and the ***imaginary parts*** as the *y*-coordinates in the ***Cartesian plane.***

Composite numbers. ***Natural numbers*** greater than one that are not ***prime numbers.*** These are numbers that can be ***factored.*** E.g., $12 = 3 \times 4$.

Countable. Term describing the ***cardinality*** of any collection of things that can be listed. E.g., the collection of ***natural numbers*** (1, 2, 3 . . .) or the ***integers*** (0, 1, −1, 2, −2 . . .).

Critical line. Vertical line in the ***complex plane*** determined by all ***complex numbers*** with a ***real part*** equal to one-half. The ***Riemann hypothesis*** is a conjecture that all ***nontrivial zeta zeros*** are ***complex numbers*** on the critical line.

Critical strip. Portion of the ***complex plane*** consisting of all ***complex numbers*** with a ***real part*** bigger than zero but less than one. It is known that all the ***nontrivial zeta zeros*** are here.

Curvature (positive or negative). Quantification of the amount of curviness in any geometric object. It can vary across the object. Positive curvature is characteristic of regions that are spherically shaped. Negative curvature is characteristic of regions that are shaped like saddles.

Denumerable. See ***countable.***

Differential equation. Equation relating the rates of change of various quantities.

Differential geometry. Geometry that takes into account minute variation.

Dimension. Number of independent numerical descriptors (parameters) needed to specify a point on the object of interest. E.g., a line is one-dimensional, and a plane is two-dimensional, as is the surface of a sphere (since latitude and longitude suffice).

Dirichlet L-series. Generalizations of ***Riemann's zeta function*** invented by Dirichlet in order to study ***primes*** in ***arithmetic progressions.***

Discrete mathematics. Generally thought of as ***combinatorics*** and discrete (as opposed to "continuous") probability theory.

Divisible, divisibility. A number *A* is divisible by *B* if *B* goes into *A* evenly. E.g., 12 is divisible by 1, 2, 3, 4, 6, and 12.

Divisor. If *A* is ***divisible*** by *B,* then *B* is a divisor of *A.*

Dynamical systems. Mathematical study of a system in motion, e.g., the individual motions of a system of interacting particles, the motion of a billiard ball on a billiard table, or even the rise and fall of competing populations.

Dyson-Montgomery-Odlyzko law. More a conjecture than a law, it states that the ***spacing distribution*** for the ***nontrivial zeta zeros*** is precisely that of the spacing distribution for the ***eigenvalues*** of a randomly chosen ***Hermitian matrix.***

e. ***Real number*** (moreover, a ***transcendental number***) whose decimal expansion starts off as 2.71828 . . . It is used as the ***base*** in most ***exponential growth laws,*** and it is the base implicit in the ***natural logarithm.***

Eigenvalue. Amount by which a ***matrix*** stretches or contracts an ***eigenvector.*** To each eigenvector there corresponds an eigenvalue.

Eigenvector. State of space transformed in "scale" by a ***matrix.*** If the state is color, then the scale is naturally considered as intensifying or dulling the color. A matrix has a specific collection of eigenvectors.

Ergodic, ergodicity. Describes a particle trajectory that comes arbitrarily close to any point in space.

Euler factorization. Manner of expressing the ***harmonic series,*** or any ***Dirichlet L-series*** (which includes the ***Riemann zeta function***) as a product of infinitely many factors. This provided the key idea for Riemann's approach to studying the accumulation of the ***primes*** through the Riemann zeta function.

Exponent. Power to which a number (the ***base***) is raised.

Exponential growth, the exponential. Rate of growth in which the amount of "stuff" accumulated after a given time is equal to a particular ***base*** raised to an ***exponent*** equal to the time elapsed. To say that something grows like the exponential usually means that the base is taken to be the number ***e.***

Exponentiation. Process of taking a power of a number.

Factor. (1) ***Divisor.*** (2) To express an integer as a product of ***integers*** different from 1 and −1.

Fermat's Last Theorem. Statement that the sum of two natural numbers raised to the same power can never equal a natural number raised to that power (e.g., the sum of two cubes is not also a cube). Proved (350 years after it was first stated) by the mathematician Andrew Wiles of Princeton University.

Fourier analysis. Mathematics of wave phenomena and periodic motion.

Fourier transform. Mathematical operation that enables a function to be expressed as a ***superposition*** of basic waves.

Frequency. Number of times per second that a regular repeating pattern moves past a given point.

Function, mathematical function. Well-defined process for turning one number into another number, or for associating one number with another number. Examples of the former are ***polynomials,*** among which is the squaring function, whereby a number (e.g., 2) is turned into its square (4). An example of the latter is the temperature in your bedroom measured at some number of seconds after midnight tonight. In this case, the number associated with 100 (i.e., the value of the bedroom temperature function at 100) would be the temperature in the room 100 seconds after midnight.

Fundamental theorem of arithmetic. Any ***natural number*** can be expressed uniquely (up to the order of appearance) as a product of ***prime factors.*** See ***prime factorization.***

Gaussian integers. Analogue of the ***integers*** for the ***complex plane*** consisting of all ***complex numbers*** whose ***real*** and ***imaginary parts*** are both integers.

Gaussian primes. Analogue of ***prime numbers*** for ***Gaussian integers.*** Thus, a Gaussian prime is such that it has no ***factors*** among the Gaussian integers other than itself, 1, −1, i, and $-i$.

Geodesic. Shortest path between two points. On a flat surface a geodesic is a Euclidean straight line. On a sphere, a geodesic is an arc of a ***great circle.***

Graph. Mathematical representation of a network.

Graph of a function. In its simplest form this is the picture drawn on the ***Cartesian plane*** in which for every point on the graph, the y-coordinate (vertical displacement) is the output of the function on the input that is the x-coordinate (horizontal displacement) of the point.

Great circle. Circumference of a sphere.

Group theory. Mathematical study of symmetry.

Hamiltonian. Matrix that encodes the dynamics within a physical system, e.g., the forces inside an atom.

Harmonic series. Infinite series obtained by adding the ***reciprocal*** of each ***natural number:*** $1 + 1/2 + 1/3 + \ldots$ It is known to diverge to infinity (i.e., the successive accumulation of reciprocals will eventually exceed any number).

Height (of a nontrivial zeta zero). According to the ***Riemann hypothesis,*** all ***nontrivial zeta zeros*** lie on the vertical ***critical line*** in the ***complex plane.*** The height of such a zeta zero is its distance from the ***real axis,*** or equivalently, its ***imaginary part.*** Also called the level of a nontrivial zeta zero.

Hermitian matrix. Matrix with ***complex number*** entries that has a symmetry in which the entry in a given row and column has the same ***real part,*** but an opposite (in sign) ***imaginary part,*** as the entry "across the diagonal," defined as the entry with row and column indices interchanged. This means that entries on the diagonal must be real numbers.

Hilbert space. Mathematical structure that generalizes our notion of two- or three-dimensional space. It can be used to describe the collection of all the possible ***wave functions*** of an atom or nucleus.

Hyperbolic geometry. Non-Euclidean geometry in which given a line and a point off the line, there are an infinity of lines through that point which do not intersect the original line. It has the property that "straight" lines which start very close to one another very quickly separate widely (e.g., the ***Poincaré disk***).

Hyperbolic triangle. Closed three-sided shape in a planar ***hyperbolic geometry*** (e.g., a triangle in the ***Poincaré disk***).

i. Notation generally used to represent the (positive) square root of minus one. In the ***complex plane*** it is the point on the ***imaginary axis*** one unit above the ***real axis.***

Imaginary axis. Vertical or y-axis in the ***complex plane.***

Imaginary numbers. Complex numbers that have a ***real part*** equal to zero. These are the complex numbers that make up the ***imaginary axis.***

Imaginary part. Any ***complex number*** can be expressed as $A + iB$, where A and B are ***real numbers*** and ***i*** is the square root of -1. B is called the imaginary part of the complex number $A + iB$.

Infinite series. Summation of an infinite sequence of numbers, achieved by adding more and more of the numbers on the list. If this process gets as close as you like to a fixed number, then the series converges to that number; if it passes any fixed number, it diverges to infinity. (If it neither diverges to infinity nor converges to a particular number, it simply diverges.)

Integer. Any of the numbers $\ldots -2, -1, 0, 1, 2 \ldots$

Integrable system. Dynamical system that does not exhibit ***chaotic*** dynamics. E.g., billiard balls moving on a rectangular billiard table.

Integral. Symbolic representation of an integration.

Integration. Technique of the ***calculus*** that enables evaluation of summations of an infinite collection of infinitesimal summands.

Irrational number. Real number that is not a ***rational number,*** i.e., that cannot be represented as a ratio of two ***integers.*** E.g., ***e*** and π are irrational numbers.

Logarithm. Used in reference to a particular ***base,*** extracts the ***exponent*** required to achieve a given number as a power of the base.

Logarithmic growth. Rate of growth in which the amount of "stuff" accumulated after a given time is equal to a particular ***logarithm*** of the time elapsed.

Logarithmic integral. Certain ***integral*** used by Gauss to estimate the accumulation of the primes.

Matrices. Plural of ***matrix.***

Matrix. Rectangular checkerboard-like grid with numerical entries, much like a spreadsheet. A matrix represents a transformation of space, physical or abstract.

Mertens conjecture. Idea that the accumulation of the ***Möbius function*** up to any given number is within the square root of that number. For example, up to 10,000, the sum of the Möbius function values is between −100 and 100. Eventually disproved in 1985 by A. Odlyzko and H. Le Ride.

Möbius function. On input of a ***natural number,*** this function gives an output of 0 if the input is divisible by a square (e.g., 4 or 12), −1 if the input is the product of an odd number of distinct ***primes*** (e.g., 5 or 105), and 1 if it is the product of an even number of distinct primes (e.g., 6 or 210). Stieltjes thought he had found a way to prove the ***Riemann hypothesis*** by analyzing the way in which these values accumulated.

Natural logarithm. Logarithm that corresponds to a ***base*** of ***e.*** The natural logarithm of a number is the power to which *e* is raised in order to obtain the number. It is a mathematical function that reflects the way we perceive things—it grows very slowly in the sense that very large numbers still have a natural logarithm which is quite small.

Natural numbers. The numbers 1, 2, 3, . . .

Non-Euclidean geometry. Generally, any geometry in which one of Euclid's five axioms fails, but usually a geometry in which the parallel postulate fails, so that given a line and a point off the line it is possible that either there are no lines through the point parallel to the original, or more than one such line. The former is true in ***spherical*** or ***elliptic geometry,*** and the latter in ***hyperbolic geometry.***

Noncommutative. Literally, "not ***commutative.***" Refers to an operation in which the order of doing things matters. For example, in multiplying ***matrix*** *A* and matrix *B,* usually $A \times B$ does not equal $B \times A$. I.e., matrix multiplication is noncommutative.

Nontrivial zeta zero. Zeta zero that is not on the ***real axis,*** i.e., a zeta zero that is not trivial. Riemann showed that these all occur inside the ***critical strip*** in the ***complex plane,*** and conjectured that they all occur on the ***critical line.*** This conjecure is the ***Riemann hypothesis*** and its resolution has implications for our understanding of the growth of the ***prime numbers*** and their distribution among the ***integers.*** See also ***trivial zeta zeros.***

Number theorist. Someone who studies ***number theory.***

Number theory. Study of the laws governing the composition and behavior of the ***integers*** and in particular the ***prime numbers.***

Operator. Mathematical transformation of a "space," which can be physical space, or abstract like a ***Hilbert space.***

Pair correlation. Measure of the interrelatedness of a collection of numbers by considering the distribution of all interpair distances.

Partial differential equation. Differential equation that incorporates the rates of change of a variety of different parameters, for example, change with respect to time and space.

Periodic function. Mathematical function whose ***graph*** is periodic—i.e., is a regularly repeating pattern.

Periodic orbit, periodic trajectory. Path that repeats itself, like that taken by a billiard ball bouncing back and forth across a billiard table.

Poincaré disk. Model of ***hyperbolic geometry,*** a ***non-Euclidean geometry*** that takes place in a circular domain in which lines are given by interior circular arcs (and diameters) perpendicular to the enclosing circular boundary.

Poisson process. Basic model of random behavior. Given a temporal cast, say the emission of a particle, it is such that the probability of the time of the next occurrence of the event is independent of the time of the last occurrence.

Polynomial. Type of mathematical function in which the output is a combination of multiples of powers of the input. In particular, if it is a combination of multiples of the zeroth, first, and second powers it is a quadratic polynomial. If it also includes the third power, it is a cubic polynomial.

Primality test. Algorithm that determines if a ***natural number*** is ***prime.***

Prime factorization. Expression of a ***natural number*** (or ***integer***) as a product of ***prime numbers*** (its "prime ***factors***"), as guaranteed by the ***fundamental theorem of arithmetic.***

Prime number. Natural number greater than one that is ***divisible*** by only itself and one.

Prime Number Theorem. Originally conjectured by Gauss, later proved by Hadamard and de la Vallée-Poussin. States that the number of ***primes*** less than a given value is ***asymptotically*** equal to that value divided by its ***natural logarithm.***

Quantization, to quantize. Mathematical process that associates a quantum system with a macroscopic system described by ***classical mechanics.***

Quantum chaos. Phenomena exhibited by ***quantum mechanical systems*** whose classical (macroscopic) analogs are chaotic. See ***chaotic dynamics, quantum mechanics.***

Quantum mechanics, quantum mechanical systems. Laws of physics on the scale of the atom and smaller. Characterized by their indeterminacy. Particles no longer have a definite location and momentum but are instead described in terms of ***wave function,*** which encodes a probability that a particle is in a given state.

Random matrix. Matrix constructed by some well-defined chancelike process, e.g., one in which every entry in the grid is chosen at random from a bell curve distribution of the numbers between 0 and 1. In relation to the ***Riemann hypothesis,*** the process is one of choosing a random matrix with some specified mathematical symmetry (e.g., ***Hermitian matrix***).

Random walk. Random process in which at each time-step a particle ("walker") has a chance of moving in one of several directions, each with a given probability.

Rational number. Number that can be represented as a fraction.

Real axis. The x-axis of the ***complex plane,*** consisting of all the ***real numbers*** (i.e., ***complex numbers*** with an ***imaginary*** part equal to zero).

Real part. Any ***complex number*** can be expressed as $A + iB$, where A and B are ***real numbers*** and ***i*** is the square root of -1. A is called the real part of the complex number $A + iB$.

Real numbers. All numbers represented by (possibly infinite) decimal expansions; the numbers comprising the ***rational*** and ***irrational numbers.***

Reciprocal. "One over" a number. I.e., the reciprocal of a number n is $1/n$.

Relative error. Given the truth and an approximation to the truth, the relative error in the approximation is the magnitude of the ratio of the difference between the truth and the approximation to the truth.

Riemann hypothesis. That the ***nontrivial zeta zeros*** (of ***the Riemann zeta function***) are ***complex numbers*** all of which are on the ***critical line*** in the ***complex plane,*** i.e., all have a ***real part*** equal to one-half. This hypothesis is equivalent to an extremely precise estimate of the distribution of the primes among the natural numbers. (Also known as Riemann's hypothesis.)

Riemann zeta function. Mathematical function cooked up by Riemann, using as a template the ***harmonic series,*** in order to study the distribution of the primes. (Also known as Riemann's zeta function.)

Riemannian geometry. Geometric theory that allows for infinitesimal local variation as well as a more general notion of distance. Provides mathematical foundations for general relativity and modern cosmology.

Semiclassical limit. Limiting procedure whereby physicists and mathematicians turn formulas that describe ***quantum mechanical systems*** into formulas describing the dynamics of macroscopic systems. See ***quantum mechanics.***

Sensitive dependence on initial conditions. Property of ***chaotic dynamical systems:*** two slightly different starting conditions (e.g., one a calm moment at a point in the atmosphere, and the other a butterfly flapping its wings at that same moment in the same place) can evolve to dramatically different outcomes (e.g., temperate weather versus a hurricane). See ***chaotic dynamics*** and ***dynamical systems.***

Sinusoid. Infinite periodic wave, characteristic of the "sine function" of trigonometry. See ***periodic function.***

Skewes number. First value for which Gauss's estimate for the prime accumulation underestimates the true value.

Spacing distribution. Given a list of numbers or points, and having derived the ***spacings,*** this is the bar chart (or "histogram") that records the percentage of spacings between various values. The distribution is usually recalibrated (normalized) so that the average spacing is one.

Spacings. Given a sequence of points on a straight line, the list of distances (spacings) between neighboring points. If these points are just points on the number line, then this is the same as taking a list of numbers, given in increasing order, and coming up with a new list of the successive differences. For example, this is how the spacings for a list of ***eigenvalues*** would be created. If these points are on a vertical line in the ***complex plane,*** like the known ***nontrivial zeta zeros,*** then this is the list of successive differences of the ***imaginary parts*** on the numbers in the list ordered according to their ***heights.***

Spectral. Referring to either the ***eigenvalues*** and ***eigenvectors*** of a ***matrix,*** or the energy levels of an atom.

Spectral theory. Study of ***matrices*** and ***operators*** and their ***eigenvalues*** and ***eigenvectors.*** See also ***matrix.***

Spectrum. (1) In a ***matrix,*** the collection of its ***eigenvalues.*** (2) In an atom, the collection of its allowable energy levels, as revealed through the process of spectroscopy. These become one and the same when considering the ***Hamiltonian matrix*** describing the dynamics within the atom.

Square-free integer. Integer that is not ***divisible*** by the square of any ***prime number.***

Superposition. With reference to wave phenomena, the adding of waves. This can be effected physically by a simultaneous sounding of two guitar strings, or any two sounds, in which case the various sound waves will be superposed atop one another, combining at the eardrum to give the sound.

Topology. Study of the properties of objects and shapes that remain unchanged during continuous transformations (i.e., morphing).

Trace. Sum of the ***eigenvalues*** of a ***matrix.***

Trace formula. Formula that relates a ***trace,*** or portions of a trace, to another summation, usually with the goal of showing that the individual ***eigenvalues*** making up the trace are related to the summands in the (assumed) related summation.

Transcendental number. Opposite of an ***algebraic number*** in that no finite collection of multiples of its powers will ever sum to zero. Examples are ***e*** and π.

Transfinite number. Number that describes the ***cardinality*** of an infinite collection of objects.

Trivial zeta zeros. Zeta zeros found on the ***real axis*** in the ***complex plane.*** They are the values $-2, -4, -6, \ldots$ (i.e., all the negative even ***integers***). So named because mathematicians feel they understand everything about these zeta zeros.

Twin prime conjecture. Conjecture that there exists an infinity of ***twin primes.***

Twin primes. Prime numbers that differ by two. E.g., 11 and 13 are twin primes.

Wave function. Mathematical ***function*** that encodes the probability that a particle in a ***quantum mechanical system*** is at a given location. See ***quantum mechanics.***

Zeta zeros. Complex number inputs for which ***Riemann's zeta function*** is equal to zero. They come in two flavors: ***trivial*** and ***nontrivial.*** The ***Riemann hypothesis*** conjectures that all the ***nontrivial zeros*** occur on the ***critical line*** in the ***complex plane.***

Further Reading and Sources

In gathering biographical materials, I've neither traveled to Egypt to walk in Euclid's sandals, nor visited Germany to sit at Riemann's desk. The biographical snapshots herein are composites built from a number of standard sources, each of which is to be recommended for further reading for the person whose mathematical interests have been piqued by this book. For information pre–World War II I've used two classic works of mathematics history: Carl Boyer's *A History of Mathematics* (2nd ed., revised by Uta C. Merzbach; New York: Wiley, 1989) and Morris Kline's three-volume masterpiece *Mathematical Thought from Ancient to Modern Times* (New York: Oxford University Press, 1972). For my money, for a historical treatment of the development of mathematics to that point, these books can't be beaten. I've also dipped into E. T. Bell's *Men of Mathematics* (New York: Simon and Schuster, 1933), a collection of avowedly romanticized biographical essays that nevertheless has served, and continues to excite the minds of, generations of budding mathematicians. In particular, Dedekind's description of Riemann's last moments can be found therein on pages 502–3.

Each of the above works is broad in scope. I've also benefited from reading some fine individual mathematical biographies: Constance Reid's magnificent *Hilbert* (New York: Springer-Verlag, 1970), William Dunham's *Euler: Master of Us All* (Washington, D.C.: Mathematical Association of America, 1999), *Jacques Hadamard: A Universal Mathematician* by V. G. Mazia and T. O. Shaposhnikova (Providence, R.I.: American Mathematical Society, 1998; this also contains some excellent material on the Prime Number Theorem), and the less technical *Alan Turing: The Enigma,* by Andrew Hodges (New York: Simon and Schuster, 1983). Other biographical data have been gleaned from the outstanding and award-winning Web site "MacTutor History of Mathematics archive": http://www-gap.dcs.st-and.ac.uk/~history/BiogIndex.html, maintained by the School of Mathematics and Statistics at the University of St. Andrews, Scotland.

My discussions of *e* and *i* have been informed by two excellent expository mathematical works: Eli Maor's *e: The Story of a Number* (Princeton, N.J.: Princeton University Press, 1994) and Paul Nahin's *An Imaginary Tale* (Princeton, N.J.: Princeton University Press, 1998). My retelling in chapter 4 of the contents of Gauss's letter to Encke uses the translation provided in a wonderful article by L. J. Goldstein, "A History of the Prime Number Theorem" (*American Mathematical Monthly* 80, no. 6 (1973), pages 599–615). Alain

Connes's comments presented in chapter 13 were found in the delightful and intriguing *Conversations on Mind, Matter, and Mathematics* by Jean-Pierre Changeux and Alain Connes (Princeton, N.J.: Princeton University Press, 1995).

I hope that this book has revealed new intellectual vistas for some readers. For those who have become intrigued by the magical and mysterious properties of the primes, Paolo Ribenboim's *The Little Book of Bigger Primes* (New York: Springer, 1996), *The New Book of Prime Number Records* (New York: Springer, 1996), and *My Numbers, My Friends* (New York: Springer, 2000) are lovely mathematical introductions to the subject. A somewhat quirky but accessible introduction to Fourier analysis (requiring little more than the courage to revisit trigonometry) is *Who Is Fourier? A Mathematical Adventure* by the Transnational College of LEX, Tokyo (Belmont, Mass.: Language Research Foundation, 1997). Another of its books—*What Is Quantum Mechanics?* (Belmont, Mass.: Language Research Foundation, 1995)—is of a similar flavor and worth a look to those who want to obtain some idea of the mathematics behind quantum mechanics. The subject of chaos received its definitive popular exposition in James Gleick's *Chaos: Making a New Science* (New York: Viking, 1987). Edward Lorenz's *The Essence of Chaos* (Seattle, Wash.: University of Washington Press, 1993) is to be read for a personal account of the origins and meaning of the butterfly effect, as well as its delightfully clear mathematical explanations.

Harold Edwards's book *Riemann's Zeta Function* (Mineola, N.Y.: Dover Publications, 2001) remains the bible for those who want the full technical history (through the early 1970s) of work related to the Riemann hypothesis. In particular, it contains a translation of Riemann's original paper containing the statement of the Riemann hypothesis. *The Prime Numbers and Their Distribution* by Gérald Tenenbaum and Michel Mendès-France (trans. Philip G. Spain; Providence, R.I.: American Mathematical Society, 2000) gives a quick mathematical introduction to the Riemann hypothesis as well as a nice discussion of probabilistic aspects of primes.

Index

About the Author

Dan Rockmore is a professor of mathematics and computer science at Dartmouth College. He lives in New Hampshire with his wife, son, and golden retriever.

A Note on the Type

This book was set in Adobe Garamond. Designed for the Adobe Corporation by Robert Slimbach, the fonts are based on types first cut by Claude Garamond (c. 1480–1561). Garamond was a pupil of Geoffroy Tory and is believed to have followed the Venetian models, although he introduced a number of important differences, and it is to him that we owe the letter we now know as "old style." He gave to his letters a certain elegance and feeling of movement that won their creator an immediate reputation and the patronage of Francis I of France.

Composed by North Market Street Graphics
Lancaster, Pennsylvania
Printed and bound by Berryville Graphics,
Berryville, Virginia
Designed by Anthea Lingeman